Chemical Thermodynamics for Metals and Materials

Chemical Thermodynamics for Metals and Materials

Hae-Geon Lee
Pohang University of Science & Technology
Korea

Imperial College Press

Published by

Imperial College Press
57 Shelton Street
Covent Garden
London WC2H 9HE

Distributed by

World Scientific Publishing Co. Pte. Ltd.
5 Toh Tuck Link, Singapore 596224
USA office: 27 Warren Street, Suite 401-402, Hackensack, NJ 07601
UK office: 57 Shelton Street, Covent Garden, London WC2H 9HE

Library of Congress Cataloging-in-Publication Data
Lee, Hae-Geon.
  Chemical thermodynamics for metals and materials / Hae-Geon Lee.
  p. cm.
  ISBN-13 978-1-86094-177-1
  ISBN-10 1-86094-177-X
  Includes index.
  1. Metals--Thermal properties. 2. Materials--Thermal properties.

TA460 .L415 1999
620.1'696--dc21

2002281277

British Library Cataloguing-in-Publication Data
A catalogue record for this book is available from the British Library.

First published 1999
Reprinted 2000, 2001, 2006, 2009, 2010

Copyright © 1999 by Imperial College Press

All rights reserved. This book, or parts thereof, may not be reproduced in any form or by any means, electronic or mechanical, including photocopying, recording or any information storage and retrieval system now known or to be invented, without written permission from the Publisher.

For photocopying of material in this volume, please pay a copying fee through the Copyright Clearance Center, Inc., 222 Rosewood Drive, Danvers, MA 01923, USA. In this case permission to photocopy is not required from the publisher.

Printed by FuIsland Offset Printing (S) Pte Ltd, Singapore

To
Myoung-Hi
and
Hanna, Rebecca and Chris

PREFACE

This book with CD-ROM presents classical thermodynamics with an emphasis on chemical aspects. It is written primarily for students and graduate engineers of metals and materials. Since its treatment of the subject is sufficiently general, students in related fields such as physical chemistry and chemical engineering can also use it.

As thermodynamics is a key discipline in most science and engineering fields, a great number of books, each claiming originality in presentation and approach, have been published on the subject. However, thermodynamics is still a confusing subject for uninitiated students and an "easy to forget" one for graduate engineers.

After many years of experience both teaching thermodynamics at university and actually practising it in industry, I have concluded that the most effective way of presenting thermodynamics is to simulate the method that a lecturer would employ in class. When teaching, the lecturer may write important equations and concepts on the board, draw underlines, circle or place tick marks to emphasise important points, draw arrows to indicate relationships, use coloured chalk for visual effect, and erase some parts to write new lines. He/she may even repeat some parts to stress their importance. A book written on paper alone cannot properly simulate the techniques mentioned above.

This package consists of a book and a computer-aided learning package, and is both unique and beneficial in that it simulates the classroom interaction much more closely due to its employment of multimedia capabilities. Unlike the passive presentation found in most textbooks, this package provides the user with an interactive learning environment. Fast topic selection, free repetition and cross-referencing by toggling between sections or even other packages are just some of the advantages this package has. This approach is in many important respects better than those adopted by other available books on the subject.

This package provides a comprehensive treatment of all the important topics of thermodynamics. It is comprised of a number of smaller sections, each of which deals with a specific topic of thermodynamics. Each section is divided into three parts:

    Text   : This part covers the fundamental concepts of thermodynamics.

    Examples : This part presents extended concepts through questions and answers.

    Exercises : This part develops skills necessary to deal with numeric problems.

This book is intended for use primarily at the undergraduate level, but will also be useful to the practising graduate engineers in industry.

Having been evolved from my teaching materials, this book unavoidably includes a blending of knowledge of many other authors with that of my own. I acknowledge their contributions. I am indebted to my teacher, Professor Y.K. Rao for introducing me to the world of thermodynamics. I am particularly grateful to my former colleague Professor Peter Hayes at The University of Queensland, Australia, for making many useful comments and giving me constant encouragement.

I want to acknowledge the major effort expended by many of my students: Y. B. Kang, T. I. Kim, C. H. Park and S. S. Lee for helping me to design this electronic book and H. J. Kong for helping to typeset the manuscript.

I am also pleased to acknowledge the financial support from The Commonwealth Government of Australia and Pohang University of Science and Technology, Korea.

Finally, I am deeply thankful to my wife and children for the love and encouragement they have given to me.

Hae-Geon Lee

CONTENTS

Preface   vii

1. Fundamental Principles and Functions   1

1.1 First Law of Thermodynamics   1
    1.1.1 Heat, Work and Internal Energy   1
    1.1.2 Enthalpy and Heat Capacity   6
    1.1.3 Enthalpy Change (ΔH)   13

1.2 Second Law of Thermodynamics   16
    1.2.1 Reversible and Irreversible Processes   16
    1.2.2 Entropy (S)   17
    1.2.3 Criterion of Equilibrium   27
    1.2.4 Heat Engines   28

1.3 Auxiliary Functions   32
    1.3.1 Free Energies   32
    1.3.2 Effect of Pressure on Free Energy   37
    1.3.3 Effect of Temperature on Free Energy   38
    1.3.4 Some Useful Equations   40

1.4 Third Law of Thermodynamics   43
    1.4.1 Third Law of Thermodynamics   43
    1.4.2 Absolute Entropies   45

1.5 Calculation of Enthalpies and Free Energies   47
    1.5.1 Standard States   47
    1.5.2 Heat of Formation   49
    1.5.3 Heat of Reaction   51
    1.5.4 Adiabatic Flame Temperature   55
    1.5.5 Gibbs Free Energy Changes   57

2. Solutions — 61

2.1 Behaviour of Gases — 61
    2.1.1 Ideal Gases — 61
    2.1.2 Fugacities and Real Gases — 64

2.2 Thermodynamic Functions of Mixing — 67
    2.2.1 Activities and Chemical Potentials — 67
    2.2.2 Partial Properties — 74

2.3 Behaviour of Solutions — 80
    2.3.1 Ideal Solutions — 80
    2.3.2 Non-ideal Solutions and Excess Properties — 83
    2.3.3 Dilute Solutions — 87
    2.3.4 Gibbs-Duhem Equation — 89
    2.3.5 Solution Models — 92

3. Equilibria — 97

3.1 Reaction Equilibria — 97
    3.1.1 Equilibrium Constant — 97
    3.1.2 Criteria of Reaction Equilibrium — 101
    3.1.3 Effect of Temperature on Equilibrium Constant — 111
    3.1.4 Effect of Pressure on Equilibrium Constant — 114
    3.1.5 Le Chatelier's Principle — 116
    3.1.6 Alternative Standard States — 117
    3.1.7 Interaction Coefficients — 125
    3.1.8 Ellingham Diagram — 127

3.2 Phase Equilibria — 131
    3.2.1 Phase Rule — 131
    3.2.2 Phase Transformations — 136
    3.2.3 Phase Equilibria and Free Energies — 144

4. Phase Diagrams — 157

4.1 Unary Systems — 157
    4.1.1 Pressure-Temperature Diagrams — 157

	4.1.2 Allotropy		163
4.2	Binary Systems		166
	4.2.1	Binary Liquid Systems	166
	4.2.2	Binary Systems without Solid Solution	177
	4.2.3	Binary Systems with Solid Solution	186
	4.2.4	Thermodynamic Models	195
4.3	Ternary Systems		204
	4.3.1	Composition Triangles	204
	4.3.2	Polythermal Projections	211
	4.3.3	Isothermal Sections	223

5. Electrochemistry — 233

5.1	Electrochemical Concepts and Thermodynamics		233
	5.1.1	Basic Electrochemical Concepts	233
	5.1.2	Electrochemical Cell Thermodynamics	236
5.2	Electrochemical Cells		243
	5.2.1	Cells and Electrodes	243
	5.2.2	Concentration Cells	251
5.3	Aqueous Solutions		260
	5.3.1	Activities in Aqueous Solutions	260
	5.3.2	Solubility Products	268

Appendices		273
I	Heats of formation, standard entropies and heat capacities	275
II	Standard free energies of formation	283
III	Properties of Selected Elements	297
IV	Standard half-cell potentials in aqueous solutions	301
Index		305

CHAPTER 1

FUNDAMENTAL PRINCIPLES AND FUNCTIONS

1.1. First Law of Thermodynamics

1.1.1. Heat, Work and Internal Energy

The First Law of Thermodynamics is really a statement of the Principle of Conservation of Energy:
  Energy can neither be created, nor destroyed.
  Energy can be transported or converted from one form to another, but cannot be either created or destroyed.
  Chemical and/or physical changes are accompanied by changes in energy.
The types of energy commonly encountered include:
  Heat energy
  Work or mechanical energy
  Electrical energy
  Chemical energy

Heat (q)

Heat flows by virtue of a temperature difference. Heat will flow until temperature gradients disappear. When heat flows, energy is transferred. The sign convention is that heat is *positive* when it flows to the system from the surroundings and *negative* when it flows from the system to the surroundings.

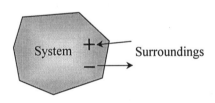

System is defined as a portion of the universe that is chosen for thermodynamic discussion and the *surroundings* is the remainder of the universe.

Work (w)

Work is the transfer of energy by interaction between the system and the surroundings. There are many types of work:
  Mechanical work
  Electrical work
  Magnetic work
  Surface tension
For now, we will be dealing mainly with mechanical work.

The system can do work on the surroundings. The surroundings can also do work on the system. The sign convention employed in this text:
  If the system does work on the surroundings, then work (*w*) is *positive*.
  If the surroundings does work on the system, then work (*w*) is *negative*.

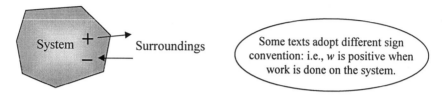

Internal Energy (U)

Energy contained in the system is called the internal energy.

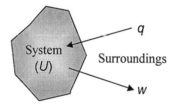

- If heat (*q*) is supplied to the system, the internal energy of the system (*U*) will increase.
- If the system does work (*w*) to the surroundings, energy will be expended and hence the internal energy (*U*) of the system will decrease.

Net change in the internal energy (ΔU) is then

$$\Delta U = q - w$$

This principle is referred to as the *First Law of Thermodynamics*: Energy may be converted from one form to another, but it cannot be created or destroyed.

Example 1

Work can be expressed in terms of a force and the displacement of its point of action. If the gas inside the cylinder shown expands and pushes the piston against the external pressure P_{ex}, can the force (F) exerted by the gas on the piston be represented by the following equation?

$$F = AP_{ex}$$

where A is the cross sectional area of the piston.

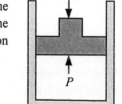

As *force* = *area* x *pressure*, the expression is correct. If the gas expands against the external pressure P_{ex} from L_1 to L_2 in the figure, the work done by the system (i.e., the gas) is

$$w = \int_{L_1}^{L_2} A P_{ex} \, dL = \int_{V_1}^{V_2} P_{ex} \, dV$$

If the external pressure is continuously adjusted so that it is kept the same as the internal pressure P,

$$w = \int_{V_1}^{V_2} P \, dV$$

Work under these conditions is called *reversible work*.

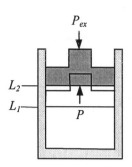

More about reversible work

Work is a mode of energy transfer which occurs due to the existence of imbalance of forces between the system and the surroundings. When the forces are *infinitesimally* unbalanced throughout the process in which energy is transferred as work, then the process is said to be reversible.

Example 2

A system can change from one state to another in many different ways. Suppose a system changes from the initial state (A) in the figure to the final state (B). Determine the work done by the system for each of the following paths:

   Path 1 : A→C→B
   Path 2 : A→E→B
   Path 3 : A→D→B

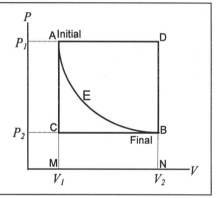

Recall that

$$w = \int_{V_1}^{V_2} P \, dV$$

Path 1: Initially the pressure is decreased from P_1 to P_2 (i.e., A→C) at the constant volume V_1 by decreasing the temperature. In this process no work has been done as there was no volume change. Next, the volume of the system expands from V_1 to V_2 (i.e., C→B) at the constant pressure P_2. The amount of work done in this process is represented by the area CBNM. This is the total work done if the system follows the path 1.

Path 2: If the system follows the path A→E→B, work done by the system is represented by the area AEBNM.

Path 3: Similarly, the amount of work done by the system is given by the area ADNM.

The amount of work done by the system depends on the path taken, and hence cannot be evaluated without a knowledge of the path.

> *Example 3*
>
> Suppose that a substance can exist in several different states as shown in the figure. The substance in state A undergoes a change to state B via state 1, and then comes back to state A via states 2 and 3.
> Is the gain of internal energy in the forward process (A→B) different from the loss in the backward process (B→A)?

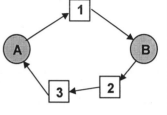

No. It should be the same. If different, the system will return to the initial state A with a net gain of internal energy. In other words, if different, the system will gain even more energy by repeating the process from nowhere. This is contrary to the First Law of Thermodynamics. Therefore the internal energy gained in the forward process must be equal to that lost in the return process.

We have seen here that internal energy (U) differs from heat (q) and work (w) in that it depends only on the state of the system, not on the path it takes. Functions which depend only on the initial and final states and not on path are called *state functions*.

> *Example 4*
>
> Which of the following thermodynamic terms are state functions?
> Temperature (T),   Pressure (P),   Heat (q),   Work (w),   Volume (V)

State functions : T, P, V Non-state functions : q, w

State functions which depend on the mass of material are called *extensive properties* (e.g., U, V). On the other hand some state functions are independent of the amount of materials. These are called *intensive properties* (e.g., P, T).

- Thermodynamics is largely concerned with the relations between state functions which characterise systems:

- A state function can be integrated between the initial (A) and final (B) states, being independent of integration path.

    (Example) $\Delta U = \int_A^B dU$

- An exact differential can be written in terms of partial derivatives. For instance, as $U = f(T, V)$,

$$dU = \left(\frac{\partial U}{\partial T}\right)_V dT + \left(\frac{\partial U}{\partial V}\right)_T dV$$

- The order of differentiation of a state function is immaterial.

$$\left[\frac{\partial}{\partial V}\left(\frac{\partial U}{\partial T}\right)_V\right]_T = \left[\frac{\partial}{\partial T}\left(\frac{\partial U}{\partial V}\right)_T\right]_V$$

- The First Law of Thermodynamics may be summarised by the following equation :

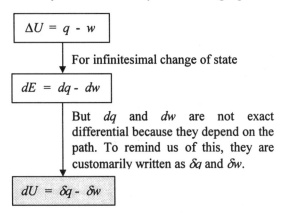

More about q and w

"Heat" and "work" are defined only for processes: heat and work are modes of energy transfer. A system cannot possess either heat or work. There is no function of state that can represent heat or work.

Exercises

1. Calculate the work done by one mole of an ideal gas when it isothermally expands from 1m³ to 10m³ at 300K.

2. A system moves from state A to state B as shown in the figure. When the system takes path 1, 500 J of heat flow into the system and 200 J of work done by the system.
   a) Calculate the change of the internal energy.
   b) If the system takes path 2, 100 J of work is done by the system. How much heat flows into the system?
   c) Now the system returns from state B to state A via path 3. 100 J of work is done on the system. Calculate the heat flow.

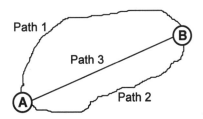

1.1.2 Enthalpy and Heat Capacity

Enthalpy (H)

If a process takes place at *constant volume*,

$$w = P\Delta V = 0$$

From the first law of thermodynamics,

$$\Delta U = q - w = q$$

Therefore, the increase or decrease in internal energy of the system is equal to the heat absorbed or released, respectively, at constant volume. If a process is carried out at a *constant pressure* rather than at a constant volume, then the work done by the system as a result of the volume change is

$$w = \int_1^2 P dV = P\int_1^2 dV = P(V_2 - V_1)$$

From the first law of thermodynamics,

$$\Delta U = q - w$$

$U_2 - U_1 \qquad\qquad P(V_2 - V_1)$

Rearrangement yields

$$(U_2 + PV_2) - (U_1 + PV_1) = q$$

The function $U + PV$ occurs frequently in chemical thermodynamics and hence it is given a special name, *enthalpy* and the symbol H.

$$H = U + PV$$

Then

$$\Delta H = H_2 - H_1 = q$$

For a system at constant pressure, therefore,

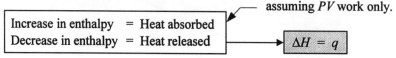

Increase in enthalpy = Heat absorbed
Decrease in enthalpy = Heat released

assuming PV work only.

$$\Delta H = q$$

For changes at other than constant pressure, ΔH still has a definite value, but $\Delta H \neq q$.

Heat Capacity (C)

The heat capacity of a system is defined as the amount of heat, q, required to raise the temperature of the system by ΔT. Thus,

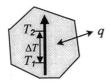

$$C = \frac{q}{\Delta T} \quad \text{where } C \text{ is the } \textit{heat capacity}.$$

For an infinitesimal change in T,

$$C = \frac{dq}{dT}$$

Heat capacity is the measure of the capacity of a system to take in energy as heat.

Recall that heat q is not a state function, so that the change in q depends on the other variable, for instance, V or P, in addition to T. Therefore,

at constant volume,
$$\delta q = dU \longrightarrow C_V = \frac{dU}{dT}$$

and,

at constant pressure,
$$\delta q = dH \longrightarrow C_P = \frac{dH}{dT}$$

where C_V is the heat capacity at constant volume, and C_P is the heat capacity at constant pressure. The variation with temperature of the heat capacity, C_P, for a substance is often given by an expression of the form :

$$C_P = a + bT + cT^{-2}$$

where a, b and c are constants to be determined empirically.

Some examples,

$$C_{P,Al_2O_3} = 106.6 + 17.8 \times 10^{-3} T - 28.5 \times 10^5 T^{-2}, J\,mol^{-1} K^{-1}$$

$$C_{P,CO_2(g)} = 44.1 + 9.04 \times 10^{-3} T - 8.54 \times 10^5 T^{-2}, J\,mol^{-1} K^{-1}$$

> **Example 1**
>
> 1) Prove the following statements :
>   a) ΔU and ΔH are usually very similar to each other for processes involving solids or liquids.
>   b) If gases are involved in a process, these may be significantly different.
> 2) If a reaction involves an increase of 1 mole of gases in the system, calculate the difference $\Delta H - \Delta U$ at 298K.

1) PV work of condensed phases is normally negligibly small :

$$\Delta H = \Delta U + \Delta(PV) \cong \Delta U.$$

If gases are involved in a process,

$$\boxed{\Delta H = \Delta U + \Delta(PV)}$$

$PV = nRT$ for a perfect gas

$$\boxed{\Delta H = \Delta U + \Delta(n)RT}$$

Therefore, if there is a change in the total number of moles of the gas phase, ΔH may be significantly different from ΔU.

2) $\Delta H - \Delta U = \Delta(n)RT = (1)(8.314 \text{ J mol}^{-1} \text{ K}^{-1})(298\text{K}) = 2.48 \text{ kJ mol}^{-1}$

> **Example 2**
>
> 1) Prove the following statements :
>   a) For condensed phases, i.e., solids or liquids, C_V and C_P are quite similar in magnitude.
>   b) C_V and C_P are significantly different for ideal gases.
> 2) C_P and C_V for argon gas are as follows:
> $C_P = 20.8 \text{ J mol}^{-1}\text{K}^{-1}$
> $C_V = 12.5 \text{ J mol}^{-1}\text{K}^{-1}$
>   Calculate $C_P - C_V$.

1) a) For condensed phases,

$$\boxed{\Delta H = \Delta U + \Delta(PV) \cong \Delta U}$$

$$C_V = \left(\frac{\partial U}{\partial T}\right)_V \qquad C_P = \left(\frac{\partial H}{\partial T}\right)_P$$

$$\boxed{C_P = C_V}$$

b) For ideal gases,

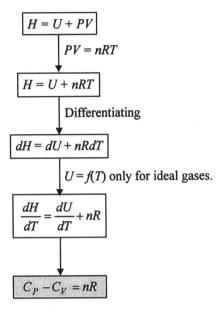

2) $C_P - C_V = 20.8 - 12.5 = 8.3$ J mol^{-1}K^{-1} : This value is very close to the gas constant R, which verifies the relationship $C_P - C_V = nR$.

Example 3

Substances usually expand with increase in temperature at constant pressure. Is C_P usually larger than C_V?

When heat is supplied to a substance,

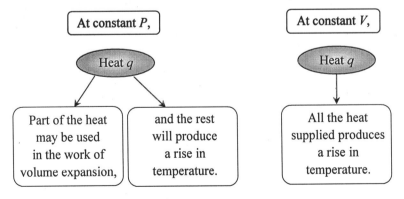

Therefore, C_p is larger than C_V.

More rigorous analysis:

$$C_P - C_V = \left(\frac{\partial H}{\partial T}\right)_P - \left(\frac{\partial U}{\partial T}\right)_V$$

from the definition of C_P and C_V.

$H = U + PV$

$$C_P - C_V = \left(\frac{\partial U}{\partial T}\right) + P\left(\frac{\partial V}{\partial T}\right)_P - \left(\frac{\partial U}{\partial T}\right)_V$$

$E = f(T, V)$

$$dU = \left(\frac{\partial U}{\partial V}\right)_T dV + \left(\frac{\partial U}{\partial T}\right)_V dT$$

$$\left(\frac{\partial U}{\partial T}\right)_P = \left(\frac{\partial U}{\partial V}\right)_T \left(\frac{\partial V}{\partial T}\right)_P + \left(\frac{\partial U}{\partial T}\right)_V$$

$$C_P - C_V = P\left(\frac{\partial V}{\partial T}\right)_P + \left(\frac{\partial U}{\partial V}\right)_T \left(\frac{\partial V}{\partial T}\right)_P$$

The contribution to C_P caused by change in the volume of the system against the constant external pressure.

The contribution from the energy required for the change in volume against the internal cohesive forces acting between the constituent particles of a substance.

For liquids and solids, which have strong internal cohesive forces, the term $(\partial U/\partial T)_P$ is large.

For gases this term is usually small compared with P. An ideal gas is a gas consisting of non-interacting particles, and hence this term is zero.

$$C_P - C_V = P\left(\frac{\partial V}{\partial T}\right)_P$$

$PV = RT$

$$C_P - C_V = R$$

Example 4

In a reversible, adiabatic process of a system comprising of one mole of an ideal gas, prove the following relationships :

$$dU = -\delta w$$

$$C_V dT = -PdV$$

1) $dU = \delta q - \delta w$, but $\delta q = 0$ in an adiabatic process.

2)

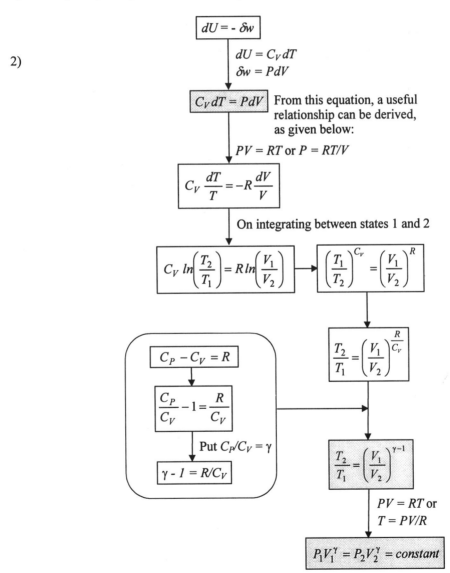

> **Example 5**
>
> For reversible adiabatic expansion of an ideal gas, we have seen
>
> $$PV^\gamma = c \text{ (constant)}$$
>
> When a system comprising of one mole of an ideal gas changes its state from (P_1, V_1, T_1) to (P_2, V_2, T_2), prove that work done by the system is
>
> $$w = \frac{P_1 V_1 - P_2 V_2}{\gamma - 1} = C_V (T_1 - T_2)$$

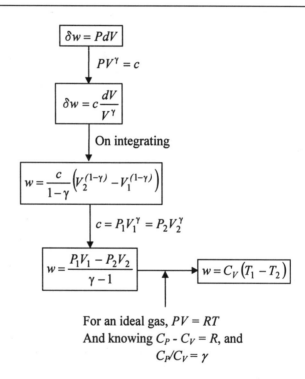

For an ideal gas, $PV = RT$
And knowing $C_P - C_V = R$, and
$C_P/C_V = \gamma$

Exercises

1. When heat is supplied to an ice-water mixture at 0°C, some of the ice melts, but the temperature remains unchanged at 0°C. What is the value of C_P of the ice-water mixture?

2. Metallic vapours generally have a monatomic constitution. According to the kinetic theory of gases, only three translational degrees of freedom need to be considered for a monatomic gas and hence the translational kinetic energy is given by

$$E = \tfrac{3}{2}nRT$$

1) Calculate ΔH when the temperature of 3 moles of the gas is raised from 700 to 1,000K.
2) Calculate C_V for the gas.
3) Calculate C_P for the gas.

1.1.3. Enthalpy Change (ΔH)

For a substance of fixed composition, the enthalpy change with change in temperature at constant pressure P can be calculated as follows :
From the definition,

$$C_P = \frac{dH}{dT} \longrightarrow \boxed{dH = C_P dT}$$

On integration from T_1 to T_2,

$$\boxed{\Delta H = \int_{T_1}^{T_2} C_P dT}$$

$$C_P = a + bT + cT^{-2}$$

$$\boxed{\Delta H = \int_{T_1}^{T_2} (a + bT + cT^{-2}) dT}$$

The enthalpy change associated with a chemical reaction or phase change at constant pressure and temperature can be calculated from the enthalpy of each species involved in the process. When species A undergoes the phase transformation from α to β,

$$A(\alpha) \xrightarrow{\text{Phase transformation}} A(\beta)$$

$$\boxed{\Delta H_t = H_{A(\beta)} - H_{A(\alpha)}}$$

The enthalpy change due to chemical reaction (ΔH) is the difference between the sum of enthalpies of the products and the sum of enthalpies of the reactants :

$$\boxed{\Delta H = \sum H_{products} - \sum H_{reactants}}$$

(Example) $Fe_2O_3 + 2Al = Al_2O_3 + 2Fe$: $\Delta H = (H_{Al_2O_3} + 2H_{Fe}) - (H_{Fe_2O_3} + 2H_{Al})$

Example 1

Pure copper melts at 1,084°C. Calculate the enthalpy change when 1 mole of copper is heated from 1,000°C to 1,100°C. ($C_{P,Cu(l)}$ = 31.4 J mol⁻¹K⁻¹, $C_{P,Cu(s)}$ = 22.6 + 6.28 x 10⁻³T, J mol⁻¹K⁻¹, Heat of fusion (ΔH_t) : 13,000J mol⁻¹)

Total enthalpy change = Enthalpy change associated with heating of solid copper to the melting temperature +
Heat of fusion at the melting temperature +
Enthalpy change associated with heating of liquid copper to the temperature of 1,100°C.

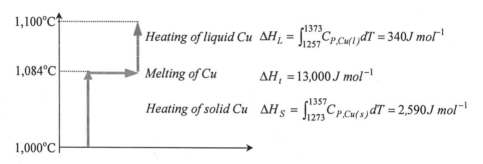

Heating of liquid Cu $\Delta H_L = \int_{1257}^{1373} C_{P,Cu(l)} dT = 340 J\ mol^{-1}$

Melting of Cu $\Delta H_t = 13,000 J\ mol^{-1}$

Heating of solid Cu $\Delta H_S = \int_{1273}^{1357} C_{P,Cu(s)} dT = 2,590 J\ mol^{-1}$

Total enthalpy change, $\Delta H = \Delta H_S + \Delta H_t + \Delta H_L = 15,930 J\ mol^{-1}$

Example 2

Molten copper is supercooled to 5°C below its true melting point (1,084°C). Nucleation of solid copper then takes place and solidification proceeds under adiabatic conditions. Calculate the percentage of the solid copper.

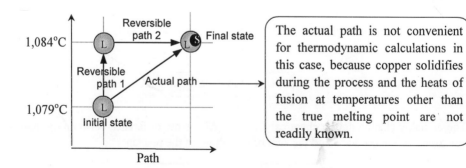

The actual path is not convenient for thermodynamic calculations in this case, because copper solidifies during the process and the heats of fusion at temperatures other than the true melting point are not readily known.

Instead, it is more convenient to take imaginary paths :
1) Liquid copper is heated from 1,079°C to 1,084°C (path 1), and then
2) portion of the liquid copper solidifies at the true melting point (path 2).

Since enthalpy is a state function, the enthalpy change along this imaginary paths should be the same as that along the true path.

$$\Delta H_{path1} = \int_{1352}^{1357} C_{P,Cu(l)} dT = 157 J\ mol^{-1}$$
$$\diagdown 31.4 J\ mol^{-1} K^{-1}$$

$$\Delta H_{path2} = -x\Delta H_t = -13{,}000x$$

where x = fraction of copper solidified, ΔH_t = heat of fusion of copper (13,000J mol^{-1})

As solidification proceeds under adiabatic conditions,

$$\Delta H_{whole\ process} = 0 = \Delta H_{path1} + \Delta H_{path2}$$

Therefore, $x = 0.012$: fraction of solid copper

Example 3

Consider the reaction between methane and oxygen to produce carbon dioxide and water vapor.

Reaction path 1 : $CH_4 + 2O_2 \rightarrow CO_2 + 2H_2O$ ΔH_1
Reaction path 2 : $CH_4 \rightarrow C + 2H_2$ ΔH_{2a}
                  $\rightarrow 2H_2 + O_2 \rightarrow 2H_2O$ ΔH_{2b}
                  $\rightarrow C + O_2 \rightarrow CO_2$ ΔH_{2c}

Prove that $\Delta H_1 = \Delta H_{2a} + \Delta H_{2b} + \Delta H_{2c}$.

Because enthalpy is a state property, the enthalpy change depends on the initial and final states only, not on the path the process follows. As the sum of all the reactions in path 2 results in the same reaction as the one in path 1, the enthalpy change should be the same for both paths.
The additive properties of enthalpy is known as *Hess's Law*. According to this law,
- the enthalpy change associated with a given chemical reaction is the same whether it takes place in one or several states, or
- enthalpies or enthalpy changes may be added or subtracted in parallel with the same manipulations performed on their respective components or reactions.

The above is in fact a different expression of the state property of enthalpy.

Exercises

1. The melting point of $CaTiSiO_5$ is 1,400°C and the heat of fusion at the normal melting point is 123,700J mol^{-1}. Calculate the heat of fusion at 1,300°C.
$C_{P,solid} = 177.4 + 23.2 \times 10^{-3} T - 40.3 \times 10^5 T^{-2}$, J mol^{-1}K^{-1}

$C_{P,liquid} = 279.6$ J mol^{-1}K^{-1}

2. Enthalpy changes resulting from temperature change can be represented on an enthalpy-temperature diagram as shown in the figure. Express on the diagram the answers to the following questions:

1) Enthalpy change when solid A melts at T_m
2) Enthalpy change when liquid A is supercooled from T_m to T_1, and then solidifies.
3) Enthalpy change when solid A is superheated from T_m to T_2 and then melts

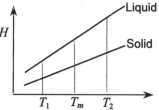

1.2. Second Law of Thermodynamics

1.2.1. Reversible and Irreversible Processes

- Wood will burn spontaneously in air if ignited, but the reverse process, i.e., the spontaneous recombination of the combustion products to wood and oxygen in air, has never been observed in nature.
- Ice at 1 atm pressure and a temperature above 0°C always melts spontaneously, but water at 1 atm pressure and a temperature above 0°C never freezes spontaneously in nature.
- Heat always flows spontaneously from higher to lower temperature systems, and never the reverse.

A process, which involves the spontaneous change of a system from a state to some other state, is called *spontaneous* or *natural process*. As such a process cannot be reversed without help of an external agency, the process is called an *irreversible process*.

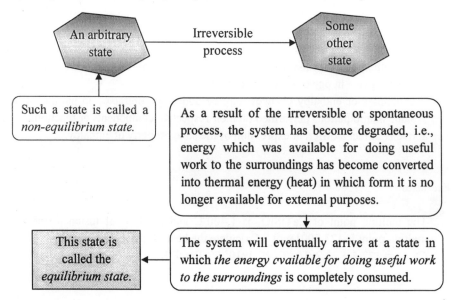

The equilibrium state is a state of rest. Once at equilibrium, a system will not move away from equilibrium unless some external agency (the surroundings) acts on it. A process during which the system is never away from equilibrium is called a *reversible process*. This statement is obviously contradictory to the definition of equilibrium.

Therefore the reversible process is an imaginary one. However, if a process proceeds with an infinitesimally small driving force in such a way that the system is never more than an infinitesimal distance from equilibrium, a condition which is virtually indistinguishable from equilibrium, then the process can be regarded as a reversible process. Thus a reversible process is infinitely slow.

1.2.2. Entropy (S)

If we are faced with the problem of deciding whether a given reaction will proceed, we might intuitively think whether there is enough energy available.

We might suppose that

if the reaction is exothermic, ⟶ it takes place spontaneously, and
($A + B \rightarrow C + D$, $\Delta H < 0$)
if the reaction is endothermic, ⟶ it does not take place spontaneously.
($A + B \rightarrow C + D$, $\Delta H > 0$)

However, there are numerous endothermic reactions that occur spontaneously. Thus energy alone then is not sufficient.

> (Example)
> Phase transformation of Sn
> Sn(white, 298K) = Sn(grey, 298K) $\Delta H = -2,100$ J mol^{-1}

Thus transformation of white tin to grey tin at 298K is exothermic, but in fact white tin is stable at 298K.

Therefore some other criterion is necessary for predicting stability or direction of a reaction. The criterion that has been found to satisfy the requirements is a quantity termed *entropy*. What is entropy then? To answer this question, we need to consider the *degradation* of a system.

There are two distinct types of spontaneous process :
- conversion of work or mechanical energy into heat, and
- flow of heat from higher to lower temperature systems.

First consider the degradation of mechanical energy of the following system:

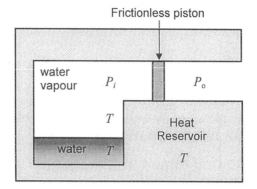

- Initially the temperature inside the cylinder equals that of the heat reservoir at T.
- The pressure inside the cylinder is the saturation vapour pressure (P_i) of water at T.
- The external pressure, P_o, is kept the same as the internal pressure P_i., $P_o = P_i$

Now the system is *at equilibrium* and the piston does not move in either direction.

- If the external pressure is suddenly decreased by dP, the piston moves out due to the pressure imbalance (Refer to the following figure).
- The volume inside the cylinder expands and the internal pressure decreases.
- Water vaporises and heat flows from the reservoir because the vaporisation is endothermic.
- After one mole of water has vaporised, the external pressure is restored to its original value, P_o and thus equilibrium is restored.

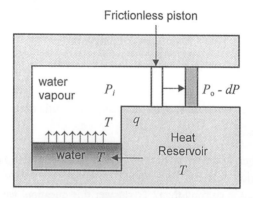

In the process described above, the system (cylinder) did work against the external pressure of P_o - dP. The amount of work done is

$$w = (P_o - dP)V$$

where V = molar volume of water vapour.

Reversible process

- The maximum work the system can do occurs when $dP \to 0$.
  $$w_{max} = P_o V$$

- When the system does the maximum work, in other words, the system undergoes a reversible process, then from the first law of thermodynamics
  $$\Delta U = q - w = q_r - w_{max}$$
  or $\quad q_r = \Delta U + w_{max}$

- q_r is the maximum amount of heat which the system can absorb from the surroundings (heat reservoir) for the vaporisation of 1 mole of water.

Irreversible process

- If the pressure drop, dP, is a finite amount, i.e., $dP \neq 0$, in other words, the system undergoes an irreversible process, then the system does less work for the same volume expansion:
  $$w = (P_o - dP)V < w_{max}$$

- Heat transferred from the surroundings to the system is
  $$q = \Delta U + w$$

Comparison

- The initial and final states are the same in both cases.
- We have seen that the maximum capacity of work that the system has is w_{max}.
- In the irreversible process, however, the amount of work the system does, w, is less than its maximum capacity.

  Where has the rest of the capacity gone?

  The mechanical energy of $(w_{max} - w)$ has been degraded to thermal energy (heat) in the system (cylinder).

In summary,

Total heat appearing in the system (q_{total})	=	Heat entering from the surroundings (q)	+	Heat produced by degradation of work due to irreversibility ($w_{max} - w$) or ($q_r - q$)

$$q_{total} = q + (q_r - q) = q_r$$

Note that the total heat is the same for both reversible and irreversible processes.

The fact that less heat enters from the surroundings in the irreversible process than in the reversible process is due to the heat produced by the degradation of work in the irreversible process.

Therefore an irreversible process is one in which the system is degraded during the process. The extent of degradation, however, differs from process to process. This suggests that there exists a quantitative measure of the extent of degradation, or degree of irreversibility, of a process.

Consider the following example in which energy is transferred by heat flow :

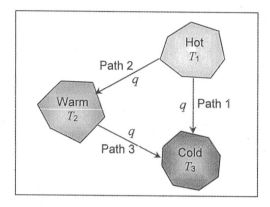

- Heat q is to flow from the hot body to the cold body.
- The flow may take place either
    (1) directly from the hot to cold bodies, i.e., Path 1, or
    (2) from the hot to warm to cold bodies, i.e., Path 2 + Path 3.
- Each of these processes is spontaneous and hence irreversible. Therefore degradation occurs in each process.

$$\boxed{\text{Path 1} = \text{Path 2} + \text{Path 3}}$$

Therefore, Degradation in Path 1 > Degradation in Path 2,
and Degradation in Path 1 > Degradation in Path 3.
Thus, Path 1 is more irreversible than either Path 2 or Path 3.

Examination of these three paths clearly indicates that the degree of irreversibility is related to temperature T and the amount of heat q.
- The more the heat flow, the higher the degree of irreversibility.
- The lower the temperature of the body to which heat flows, the higher the degree of irreversibility.

Therefore,

$$\boxed{\text{Degree of irreversibility} \propto \frac{q}{T}}$$

Fundamental Principles and Functions

Now we define a new thermodynamic function, S, called *entropy*, as

$$\text{Entropy change} = \frac{\text{Total heat input}}{\text{Temperature}}$$

or

$$\boxed{\Delta S = \frac{q_{total}}{T}}$$

The entropy change (ΔS) is the measure of the degree of irreversibility of a process.

The total entropy change associated with the process consists of two terms:

- Entropy change of the system : ΔS_{sys}
- Entropy change of the surroundings : ΔS_{sur}

$$\boxed{\Delta S_{tot} = \Delta S_{sys} + \Delta S_{sur}}$$

Recall that the total heat appearing in the system is the sum of heat entering from the surroundings (q) and heat produced by degradation due to irreversibility ($q_r - q$):

$$\boxed{q_{tot,\ system} = q_r}$$
$$\downarrow$$
$$\boxed{\Delta S_{sys} = \frac{q_r}{T}}$$

The total heat leaving the surroundings is q:

$$\boxed{q_{tot,\ sur} = -q}$$
$$\downarrow$$
$$\boxed{\Delta S_{sur} = \frac{-q}{T}}$$

Negative sign is due to heat loss of the surroundings.

$$\Delta S_{tot} = \Delta S_{sys} + \Delta S_{sur} = \frac{q_r}{T} + \frac{-q}{T} = \frac{q_r - q}{T}$$

$$\Delta S_{tot} = \frac{q_r - q}{T}$$

Reversible Process

$q = q_r$

$$\Delta S_{sys} = \frac{q_r}{T} = \frac{q}{T}$$

$$\Delta S_{sur} = \frac{-q_r}{T} = \frac{-q}{T}$$

$$\Delta S_{tot} = \frac{q_r - q}{T} = 0$$

Irreversible Process

$q < q_r$

$$\Delta S_{sys} = \frac{q_r}{T} > \frac{q}{T}$$

$$\Delta S_{sur} = \frac{-q}{T}$$

$$\Delta S_{tot} = \frac{q_r - q}{T} > 0$$

For infinitesimal change

Reversible Process

$\delta q = \delta q_r$

$$dS_{sys} = \frac{\delta q_r}{T} = \frac{\delta q}{T}$$

$$dS_{sur} = \frac{-\delta q_r}{T} = \frac{-\delta q}{T}$$

$$dS_{tot} = \frac{\delta q_r - \delta q}{T} = 0$$

Irreversible Process

$\delta q < \delta q_r$

$$dS_{sys} = \frac{\delta q_r}{T} > \frac{\delta q}{T}$$

$$dS_{sur} = \frac{-\delta q}{T}$$

$$dS_{tot} = \frac{\delta q_r - \delta q}{T} > 0$$

In the reversible process the total entropy change is zero; i.e., the entropy gain of the system is equal to the entropy loss of the surroundings, or vice versa. In other words there is no creation of entropy. Entropy is only transferred between the system and the surroundings.

In the irreversible process the total entropy change is always positive; i.e., there is a net creation of entropy. Degradation of work due to its irreversible nature accounts for this creation of entropy.

- Entropy, like energy, is a fundamental thermodynamic concept.
- Entropy is not a thing.
- Entropy is not directly measurable.

- But entropy changes are calculated from measurable quantities such as temperature, pressure, volume and heat capacity.
- Entropy is a measure of "mixed-up-ness" of a system (J.W. Gibbs).
- Entropy increases in all irreversible, spontaneous or natural processes.
- Change in entropy measures the degree of irreversibility of a process, or the capacity for spontaneous change of a process.
- Every physical or chemical process in nature takes place in such a way as to increase the sum of entropies of all the bodies taking any part in the process (Max Planck).
- Things left to themselves proceed to a state of maximum possible disorder (i.e., entropy).
- Heat always flows spontaneously from higher to lower temperature systems.
- The energy of the universe is constant, but the entropy is continually increasing until it reaches its maximum value.

The last four are some of many equivalent statements of the Second Law of Thermodynamics. Unlike internal energy (U) or enthalpy (H), entropy has its absolute value.

$$dS = \frac{\delta q}{T}$$

$\delta q = dH$ at constant pressure
$dH = C_P dT$

$$dS = \frac{C_P}{T} dT$$

Integration between the temperatures of 0K to TK yields,

$$S_T - S_0 = \int_0^T \frac{C_P}{T} dT$$

$S_0 = 0$
Refer to the 3rd law of thermodynamics

$$S_T^o = \int_0^T \frac{C_P}{T} dT$$

Superscript "^{o}" denotes the standard state at 1 atm.

Example 1

Entropy is defined as

$$dS \geq \frac{\delta q}{T}$$

for a system, where equality sign holds for reversible process and inequality sign holds for irreversible process. We know that δq is not a state function, i.e., it depends on the path the process takes. Is entropy then a state function or a non-state function?

Consider one mole of a perfect gas.

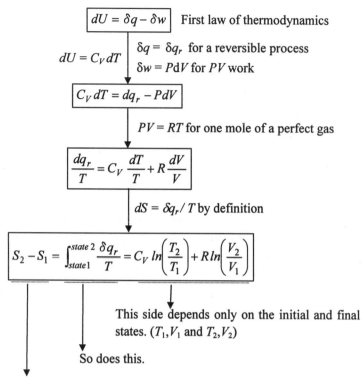

$dU = \delta q - \delta w$ First law of thermodynamics

$\delta q = \delta q_r$ for a reversible process
$\delta w = PdV$ for PV work

$dU = C_V dT$

$C_V dT = dq_r - PdV$

$PV = RT$ for one mole of a perfect gas

$$\frac{dq_r}{T} = C_V \frac{dT}{T} + R \frac{dV}{V}$$

$dS = \delta q_r / T$ by definition

$$S_2 - S_1 = \int_{state1}^{state2} \frac{\delta q_r}{T} = C_V \ln\left(\frac{T_2}{T_1}\right) + R \ln\left(\frac{V_2}{V_1}\right)$$

This side depends only on the initial and final states. (T_1, V_1 and T_2, V_2)

So does this.

Therefore S is a state function for a perfect gas.
More general arguments of this type enable us to show that entropy is a state function for all substances.

Example 2

If one mole of a perfect gas undergoes an isothermal process, is ΔS independent of pressure?

From Example 1,

$$\Delta S = C_V \ln\left(\frac{T_2}{T_1}\right) + R \ln\left(\frac{V_2}{V_1}\right)$$

$T_1 = T_2 =$ constant,
$PV = RT =$ constant

$$\Delta S = R \ln\left(\frac{P_1}{P_2}\right)$$

The entropy change depends on the pressure change. This equation tells us that as the pressure is raised the entropy of a gas decreases: i.e., molecules in the gas become more ordered.

Example 3

When a system undergoes a process at a constant pressure, does the entropy change depend on the temperature?

As entropy is a state function, we are free to choose a path from the initial and final states. The path along which a process takes place *reversibly* would be most convenient for thermodynamic calculations, because the heat absorbed or released can be directly related to the entropy change :

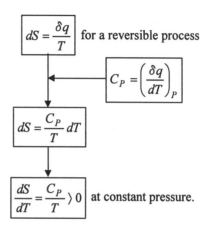

This equation tells us that the entropy of a substance held at constant pressure increases when the temperature increases.

Example 4

Suppose that energy is transferred spontaneously as heat q from a system at a fixed temperature T_1 to a system at a fixed temperature T_2 without performing any work. Prove that the total entropy change of the process, ΔS_{total}, is positive.

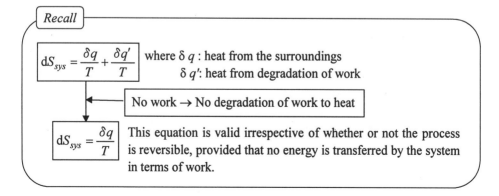

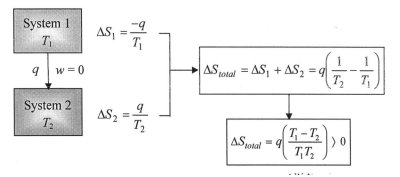

Whenever a system undergoes a change in state and the amount of work done by the system is less than the maximum possible amount of work, then there is a net entropy production.

Example 5

Prove the following statement :
An isothermal change in phase (phase transformation) of a substance produced by input of energy as heat always leads to an increase in the entropy of the substance.

$\Delta S = \dfrac{q}{T}$ for a reversible process.

$q = \Delta H_t$ (heat of transformation)
$T = T_t$ (transformation temperature)

$\Delta S = \dfrac{\Delta H_t}{T_t} \rangle 0$

Exercises

1. Calculate the work done by the reversible, isothermal expansion of 3 moles of an ideal gas from 100 litres initial volume to 300 litres final volume.

2. Tin (Sn) transforms from grey to white tin at 286K. The heat of transformation (ΔH_t) has been measured as 2.1 kJ mol^{-1}.

$$Sn(grey, 268K) \rightarrow Sn(white, 268K)$$

   1) Calculate the entropy change of the system (Tin).
   2) Calculate the entropy change of the surroundings.
   3) Calculate the total entropy change of the universe (system + surroundings).

3. Suppose that the following reaction takes place at 298K :

$$Fe_2O_3 + Al = 2Fe + Al_2O_3$$

The temperatures of both the system and the surroundings are maintained at 298K. Calculate the entropy change of the system associated with the reaction. The following data are given :

	Al	Fe	Fe_2O_3	Al_2O_3
$S°_{298}$ (J mol^{-1}K^{-1})	28.3	27.2	87.5	51.1

4. One mole of metal block at 1,000K is placed in a hot reservoir at 1,200K. The metal block eventually attains the temperature of the reservoir. Calculate the total entropy change of both the system (the metal block) and the surroundings (the reservoir). The heat capacity of the metal is given as

$$C_P = 23 + 6.3 \times 10^{-3}T, \text{ J mol}^{-1}\text{K}^{-1}$$

5. Liquid metal can be supercooled to temperatures considerably below their normal solidification temperatures. Solidification of such liquids takes place spontaneously, i.e., irreversibly. Now one mole of silver supercooled to 940°C is allowed to solidify at the same temperature. Calculate the entropy change of the system (silver). The following data are given :

$C_{P(l)} = 30.5$ J mol^{-1}K^{-1}
$C_{P(s)} = 21.3 + 8.54 \times 10^{-3}T + 1.51 \times 10^5 T^{-2}$, J mol^{-1}K^{-1}
$\Delta H_m° = 11,090$ J mol (Heat of fusion at $T_m = 961°C$)

Calculate the entropy change of the surroundings.
Does the process proceed spontaneously?

1.2.3. Criterion of Equilibrium

When a system is left to itself, it would either

remain unchanged in its initial state,	or	move spontaneously to some other state.
If this is the case, the initial state is the *equilibrium* state.		If this is the case, the system is initially in a *non-equilibrium* state. The system will spontaneously move to the equilibrium state.

- All real processes involve some degree of irreversibility and thus all real processes lead to *an increase in the total entropy* ($\Delta S_{tot} = \Delta S_{sys} + \Delta S_{sur}$).
- Entropy is not conserved, except in the hypothetical limiting case of a reversible process.
- The total entropy can never decrease. It can only increase.
- A process will cease to proceed further, if the total entropy has reached its maximum.

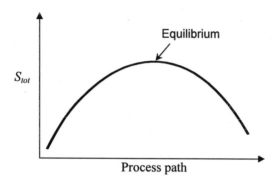

1.2.4. Heat Engines

A heat engine is a device for converting heat into work. (e.g., steam engine, internal combustion engine)
The following figure is the schematic representation of a heat engine.

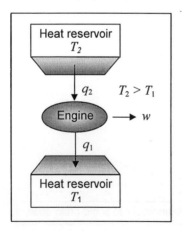

- Each cycle takes heat from the high temperature reservoir : q_2,
- use some of it to generate work : w,
- and rejects unused heat to the cold reservoir : q_1.

What is the maximum amount of work that can be obtained per cycle from the heat engine?

The maximum work we can obtain from the operation of a heat engine is that generated when all the processes are reversible, since there is no degradation of work in a reversible process.

The *Carnot cycle* is the operation of an idealised (reversible) engine in which heat transferred from a hot reservoir, is partly converted into work, and partly discarded to a cold reservoir.

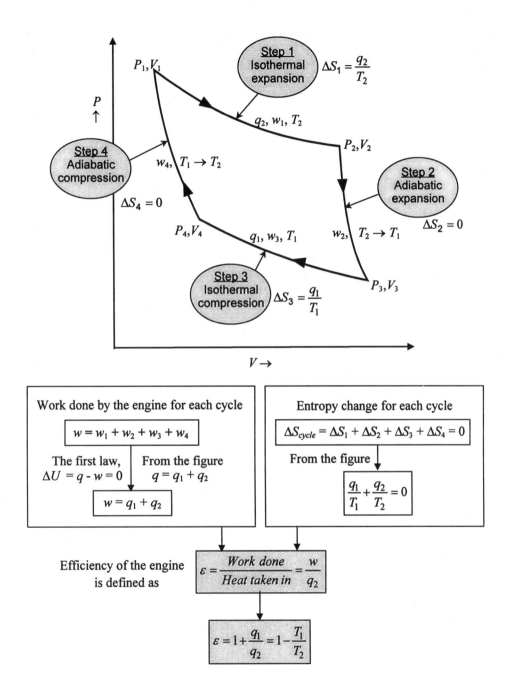

This is a remarkable result :

    The efficiency depends only on the temperatures of the reservoirs, and is independent of the nature of the engine, working substance, or the type of work performed.

Example 1

If any part of the operation of the Carnot engine is irreversible, prove that the following relationship holds:

$$\varepsilon < 1 - \frac{T_1}{T_2}$$

Entropy is a state function and hence is independent of the path a system takes.

$$\Delta S_{cycle} = \Delta S_1 + \Delta S_2$$

for a complete cycle

$$\Delta S_{cycle} = \Delta S_1 + \Delta S_2 = 0$$

$$\Delta S_1 \geq \frac{q_1}{T_1}$$

From the first law,

$$\Delta S_2 \geq \frac{q_2}{T_2}$$

$$\Delta U = (q_1 + q_2) - w = 0$$

$$\varepsilon = \frac{w}{q_2} < 1 - \frac{T_1}{T_2}$$

Example 2

A Carnot engine can be run in reverse and used to transfer energy as heat from a low temperature reservoir to a high temperature reservoir. The device is called *a heat pump*, if it is used as a heat source, or *a refrigerator*, if it is used to remove heat. Prove that in the latter case *work must be done on the engine*.

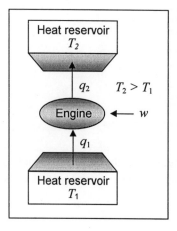

Just as for a Carnot engine running in the forward direction we have an engine running in the reverse direction.

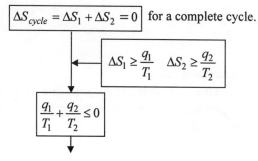

$$\Delta S_{cycle} = \Delta S_1 + \Delta S_2 = 0$$ for a complete cycle.

$$\Delta S_1 \geq \frac{q_1}{T_1} \quad \Delta S_2 \geq \frac{q_2}{T_2}$$

$$\frac{q_1}{T_1} + \frac{q_2}{T_2} \leq 0$$

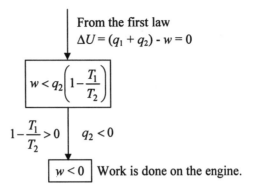

The *coefficient of performance of a heat pump* is defined as

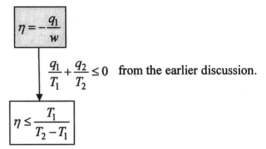

$$\eta \leq \frac{T_1}{T_2 - T_1}$$

Note that the coefficient of performance of a heat pump of a refrigerator, unlike the efficiency of a heat engine, can be greater than unity.

Exercises

1. Refer to the diagram in the text of the Carnot cycle. An engine operates between 1,200°C(T_2) and 200°C(T_1), and Step 1 (isothermal expansion) involves an expansion from $6 \times 10^5 \text{N m}^{-2}$ to $4 \times 10^4 \text{N m}^{-2}$. The working substance is one mole of an ideal gas.
   1) Calculate the efficiency of the heat engine.
   2) Calculate heat absorbed in Step 1.
   3) Calculate the amount of heat rejected in Step 3 (isothermal compression).

2. The following diagram shows the operation cycle of a Carnot refrigerator. The refrigerator operates between 25°C(T_2) and -10°C(T_1) and step 2 involves heat absorption of 500J.
   1) Calculate the coefficient of the refrigerator.
   2) Calculate the total work done per cycle.

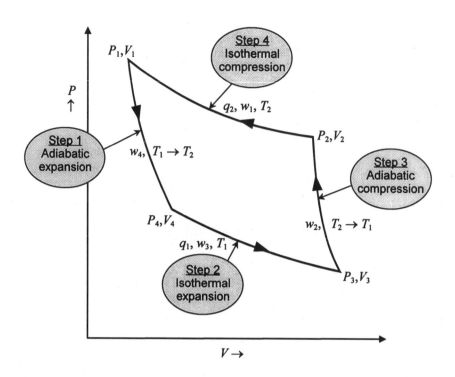

1.3. Auxiliary Functions

1.3.1. Free Energies

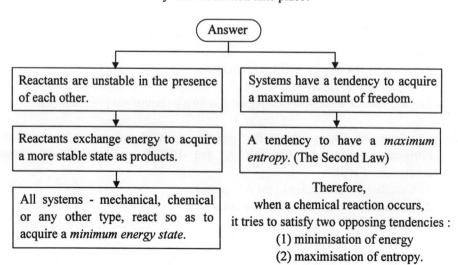

Is there a function which can represent the balance between these two opposing factors, and thus be a measure of the tendency for a reaction to take place? Such a function does exist and is known as the *free energy function* or simply *free energy*. We now derive two different free energies, namely, *Helmholtz free energy* and *Gibbs free energy*.

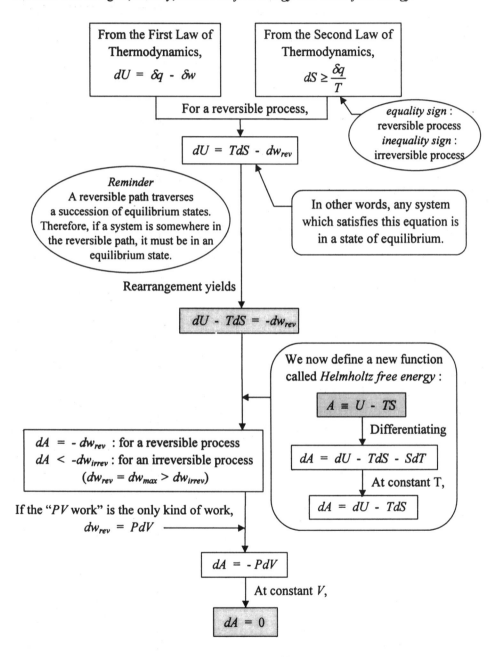

In other words, when only PV work is permitted, and T and V are kept constant, then $dA = 0$ or A is a minimum at equilibrium. Therefore, the Helmholtz free energy (A) offers criteria for thermodynamic equilibrium at constant temperature and volume. However, we are often more interested in systems at constant pressure and temperature rather than at constant volume and temperature, since most practical chemical processes take place at constant pressure. A free energy function, which is based on the constant pressure and temperature, would thus be more convenient to use. Recall that enthalpy H represents heat of reaction at constant pressure;

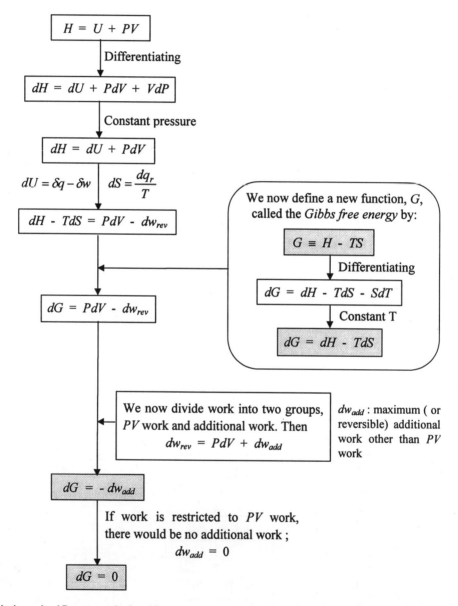

This is a significant result that if a system at constant P and T does no work other than PV work, the condition of equilibrium is that $dG = 0$ or G is minimum.

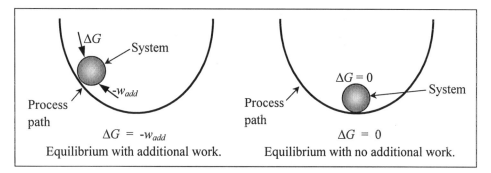
Equilibrium with additional work.   Equilibrium with no additional work.

We now have found a function G

$$dG = dH - TdS$$

which represents the balance at constant P and T between the tendencies of a system to maximise its entropy and to minimise its energy, i.e., enthalpy at constant P.

Example 1

Is the following statement true?
Change in Gibbs free energy is a measure of the net work, i.e., work other than PV work, done on the system in a reversible process at constant P and T.

$$dG = -dw_{add} \quad \xrightarrow{\text{For a finite change}} \quad \Delta G = -w_{add}$$

Additional work other than
PV work, or
net work, or
useful work.

Example 2

Is the following statement true?
The sum of the entropy change of *the system and surroundings* is zero for a reversible process and is greater that zero for a spontaneous process. Therefore one must calculate S for both the system and the surroundings to find whether a process is at equilibrium or not. However, if the Gibbs free energy function (G) is used instead, one need to calculate the Gibbs free energy change of the system (ΔG_{sys}) only, because ΔG_{sys} alone is sufficient to indicate whether or not a given process is potentially spontaneous.

According to the Second Law of Thermodynamics,

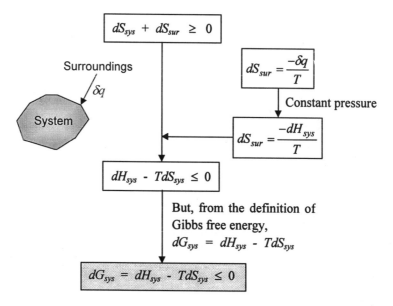

Therefore, the Gibbs free energy change of the system alone is sufficient to indicate whether a process is spontaneous ($dG_{sys} < 0$), at equilibrium ($dG_{sys} = 0$), or can not proceed ($dG_{sys} > 0$).

Example 3

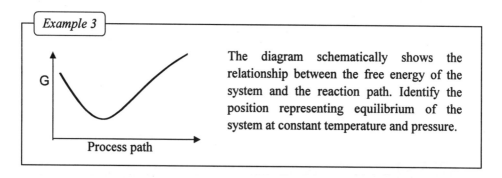

The diagram schematically shows the relationship between the free energy of the system and the reaction path. Identify the position representing equilibrium of the system at constant temperature and pressure.

Recall that $dG \leq 0$. At equilibrium $dG = 0$; i.e., the minimum point of the curve. A system always moves toward equilibrium and never away from the equilibrium state unless an external agency (the surroundings) does work on the system.

Exercises

1. Tin transforms from grey to white tin at 286K and constant pressure of 1 atm. The following thermodynamic data are given :

$C_P = 21.6 + 18.2 \times 10^{-3} T$, J mol^{-1}K^{-1} for both grey and white tin
Heat of transformation (ΔH_t°) = 2.1 J mol^{-1} at 286K

1) Calculate the free energy change (ΔG) of the transformation at 286K and 1atm.
2) Calculate ΔG of the transformation at 293K and 1atm.
3) Is the transformation at 293K spontaneous?

1.3.2. Effect of Pressure on Free Energy

From the definitions of the Gibbs free energy and enthalpy,

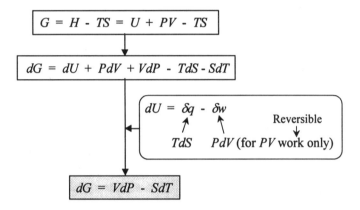

This is an important equation as it tells how free energy, and hence equilibrium position, varies with pressure and temperature.
At constant temperature,

$$dG = VdP \quad \text{or} \quad \left(\frac{\partial G}{\partial P}\right)_T = V$$

For an isothermal change from state 1 to state 2,

$$\Delta G = G_2 - G_1 = \int_{state\,1}^{state\,2} VdP$$

If the variation of V with P is known for the substance of interest, this equation can be integrated. For a simple case of one mole of an ideal gas, $PV = RT$. Thus

$$\Delta G = RT \ln\left(\frac{P_2}{P_1}\right)$$

This equation gives the change in free energy on expansion or compression.

In an isothermal process for an ideal gas,

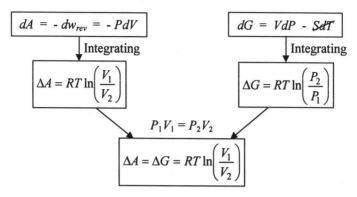

Exercises

1. One mole of an ideal gas is compressed isothermally at 298K to twice its original pressure. Calculate the change in the Gibbs free energy.

1.3.3. Effect of Temperature on Free Energy

Recall the basic equation:

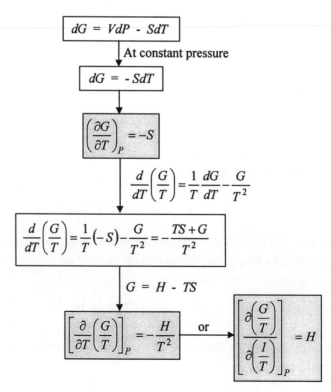

The equations given above in the shaded boxes are called the *Gibbs-Helmholtz equations*. These permit us to calculate the change in enthalpy ΔH and entropy ΔS from a knowledge of ΔG. They relate the temperature dependence of free energy, and hence the position of equilibrium to the enthalpy change.

Example 1

Internal energy (U), enthalpy (H), entropy (S), Helmholtz free energy (A) and Gibbs free energy (G) are functions of state. Each of these can be expressed as a function of two state variables. Prove the following relationships:

$$\left(\frac{\partial G}{\partial T}\right)_P = -S \qquad \left(\frac{\partial G}{\partial P}\right)_T = V \qquad \left(\frac{\partial H}{\partial P}\right)_S = V \qquad \left(\frac{\partial H}{\partial S}\right)_P = T$$

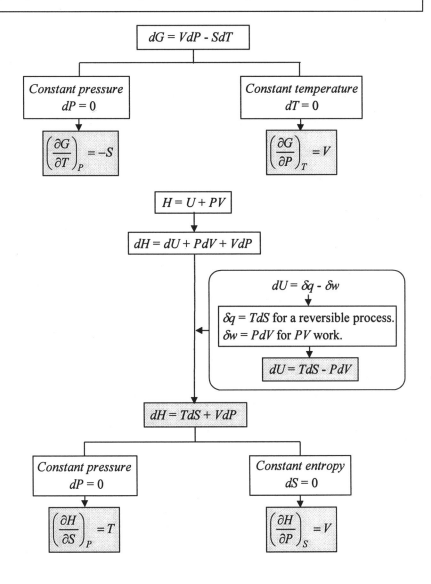

Exercises

1. The following equation shows the temperature-dependence of the free energy change of a reaction :

$$\Delta G = -1{,}750{,}000 - 15.7T \log T + 370T, \ \text{J mol}^{-1}$$

   1) Calculate ΔS for the reaction at 500K.
   2) Calculate ΔH for the reaction at 500K
   3) Will the reaction take place spontaneously at 500K?

1.3.4. Some Useful Equations

Consider the following equation :

$$\boxed{dU = \delta q - \delta w}$$

$\delta q = TdS$ and $\delta w = PdV$ for reversible process.

$$\boxed{dU = TdS - PdV}$$

constant volume:
$$\left(\frac{\partial U}{\partial S}\right)_V = T$$

constant entropy:
$$\left(\frac{\partial U}{\partial V}\right)_S = -P$$

We know the following general relationship :

$$\frac{\partial}{\partial z}\left(\frac{\partial x}{\partial y}\right)_z = \frac{\partial}{\partial y}\left(\frac{\partial x}{\partial z}\right)_y = \frac{\partial^2 x}{\partial y \partial z}$$

Thus

$$\frac{\partial}{\partial V}\left(\frac{\partial U}{\partial S}\right)_V = \left(\frac{\partial T}{\partial V}\right)_S \qquad \frac{\partial}{\partial S}\left(\frac{\partial U}{\partial V}\right)_S = -\left(\frac{\partial P}{\partial S}\right)_V$$

$$\left(\frac{\partial T}{\partial V}\right)_S = -\left(\frac{\partial P}{\partial S}\right)_V$$

Similarly,

Fundamental Principles and Functions

$dH = TdS + VdP$ → $\left(\dfrac{\partial T}{\partial P}\right)_S = \left(\dfrac{\partial V}{\partial S}\right)_P$

$dA = -PdV - SdT$ → $\left(\dfrac{\partial S}{\partial V}\right)_T = \left(\dfrac{\partial P}{\partial T}\right)_V$

$dG = VdP - SdT$ → $\left(\dfrac{\partial V}{\partial T}\right)_P = -\left(\dfrac{\partial S}{\partial P}\right)_T$

> *An example of the use of these equations*
> $(\partial S/\partial V)_T$ and $(\partial S/\partial P)_T$ are difficult to obtain in direct measurements through experiment, but may be calculated from a knowledge of the variation of P with T at constant V, $(\partial P/\partial T)_V$, and the variation of V with T at constant P, $(\partial V/\partial T)_P$.

These equations are applicable under reversible conditions, and very useful in manipulation of thermodynamic quantities. These equations are known as *Maxwell's equations*.

Example 1

Is the following statement true?
The internal energy of an ideal gas at constant temperature is independent of the volume of the gas.

For a closed system,

$$dU = TdS - PdV$$

Differentiating with respect to V at constant T,

$$\left(\dfrac{\partial U}{\partial V}\right)_T = T\left(\dfrac{\partial S}{\partial V}\right)_T - P$$

From Maxwell's equations

$$\left(\dfrac{\partial S}{\partial V}\right)_T = \left(\dfrac{\partial P}{\partial T}\right)_V$$

$$\left(\dfrac{\partial U}{\partial V}\right)_T = T\left(\dfrac{\partial P}{\partial T}\right)_V - P$$

$PV = RT$ for an ideal gas

$$\left(\dfrac{\partial P}{\partial T}\right)_V = \dfrac{R}{V} = \dfrac{P}{T}$$

$$\left(\dfrac{\partial U}{\partial V}\right)_T = 0$$

Thus, the statement is true; i.e., the internal energy of an ideal gas is independent of the volume of the gas at constant temperature. In a similar way, one may prove the following statement : The enthalpy of an ideal gas is independent of the pressure of the gas.

Example 2

The volume of thermal expansion coefficient α of a substance is defined as

$$\alpha = \frac{1}{V}\left(\frac{\partial V}{\partial T}\right)_P$$

and the compressibility β is defined as

$$\beta = -\frac{1}{V}\left(\frac{\partial V}{\partial P}\right)_T$$

Prove the following relationship :

$$C_P = C_V + \frac{\alpha^2 VT}{\beta}$$

Recall the following two equations :

$$C_P - C_V = \left[P + \left(\frac{\partial U}{\partial V}\right)_T\right]\left(\frac{\partial V}{\partial T}\right)_P \qquad \left(\frac{\partial U}{\partial V}\right)_T = T\left(\frac{\partial P}{\partial T}\right)_V - P$$

Combination of these two equations yields

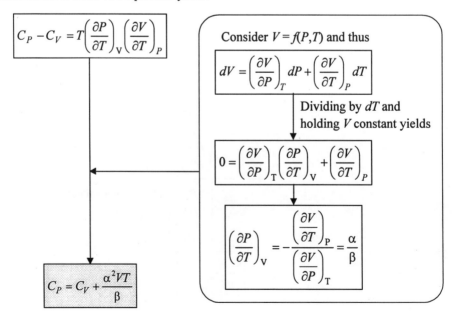

$$C_P - C_V = T\left(\frac{\partial P}{\partial T}\right)_V \left(\frac{\partial V}{\partial T}\right)_P$$

Consider $V = f(P,T)$ and thus

$$dV = \left(\frac{\partial V}{\partial P}\right)_T dP + \left(\frac{\partial V}{\partial T}\right)_P dT$$

Dividing by dT and holding V constant yields

$$0 = \left(\frac{\partial V}{\partial P}\right)_T\left(\frac{\partial P}{\partial T}\right)_V + \left(\frac{\partial V}{\partial T}\right)_P$$

$$\left(\frac{\partial P}{\partial T}\right)_V = -\frac{\left(\frac{\partial V}{\partial T}\right)_P}{\left(\frac{\partial V}{\partial P}\right)_T} = \frac{\alpha}{\beta}$$

$$C_P = C_V + \frac{\alpha^2 VT}{\beta}$$

This equation shows that C_V can be obtained from C_P, compressibility and thermal expansion coefficient for a substance. For solids, C_V is generally more difficult to measure experimentally and the equation offers a way to overcome this difficulty.

1.4. Third Law of Thermodynamics

1.4.1. Third Law of Thermodynamics

If ΔG and ΔH for a reaction are plotted as a function of temperature, results are like those shown in the following figure :

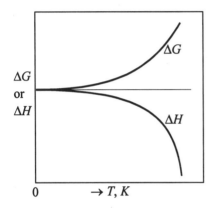

The plots suggest that as the temperature decreases, ΔG and ΔH approach equality, and ΔG approaches ΔH in magnitude with decreasing T at a much faster rate than T approaches zero.

From the definition of G and for an isothermal process,

$$\Delta G = \Delta H - T\Delta S$$

From the above we can conclude that

$$\lim_{T \to 0} \Delta S = 0$$

This equation is known as *the Third Law of Thermodynamics*.

When the temperature approaches zero, it is seen from the figure that

$$\left(\frac{\partial \Delta G}{\partial T}\right)_P = \left(\frac{\partial \Delta H}{\partial T}\right)_P = 0$$

Recall that

$$\left(\frac{\partial \Delta G}{\partial T}\right)_P = -S \quad \text{and} \quad \left(\frac{\partial \Delta H}{\partial T}\right)_P = \Delta C_P$$

Thus

$$\text{As } T \to 0 \text{ K}, \ \Delta S \to 0 \text{ and } \Delta C_P \to 0.$$

- The third law is different from the first two laws :
  1) The third law does not give rise to a new thermodynamic function. (1st law → U, 2nd law → S)
  2) The third law applies only at $T \rightarrow 0$, whereas the first two laws hold at all temperatures.
- What the third law does say is that the entropy change ΔS for any isothermal thermodynamic process becomes zero in the limit of 0 K.

Example 1

Prove the following statements :
The third law does not give rise to a new thermodynamic function.
(1st law → U, and 2nd law → S)
The third law applies only at $T \rightarrow 0$, whereas the first two laws hold at all temperatures.

What the third law does say is that the entropy change ΔS for any isothermal thermodynamic process becomes zero in the limit of 0K.

Example 2

Is the following true?
For the reaction

$$2Ag(s) + PbSO_4 = Pb(s) + Ag_2SO_4(s)$$

(Sum of the entropy of the products) - (Sum of the entropy of the reactants) = 0 at 0K.

$$\Delta S_{reaction} = \sum (S_{products}) - \sum (S_{reactants})$$

According to the third law, $\Delta S_{reaction} \rightarrow 0$ as $T \rightarrow 0$. In the limit at $T = 0K$, $\Delta S_{reaction} = 0$

Example 3

Consider the phase change of metallic tin :

$$Sn(s, grey) = Sn(s, white)$$

The two phases are in equilibrium at 13.2°C and 1 atm. The entropy change ΔS for the transformation of grey to white tin is 7.82 J mol^{-1}K^{-1}. Find the entropy change for the transformation at 0K.

According to the third law, in the limit of absolute zero there is no entropy difference between grey and white tin. The value of the entropy difference decreases from 7.82 Jmol-1K-1 to zero as the temperature decreases from 286.4K to 0K.

Exercises

1. Sulphur has two solid allotropes: Monoclinic sulphur can readily be supercooled to very low temperatures, completely bypassing the phase transformation at 368.5K. The temperature dependence of the heat capacities of both allotropes can be determined experimentally. It has been found that

$$\Delta S_{rhom} = \int_0^{368.5} C_{P(rhom)} d\ln T = 36.86 \, JK^{-1}$$

$$\Delta S_{mono} = \int_0^{368.5} C_{P(mono)} d\ln T = 37.82 \, JK^{-1}$$

$$\Delta H_t = 402 \, J$$

K
Monoclinic
368.5 — Phase transformation
Rhombohedral
0

Calculate the entropy change at 0K.

2. Calculate the entropy change for the dissociation reaction at 0K :

$$2CuO(s) = Cu_2O(s) + \tfrac{1}{2}O_2(g)$$

The following data are given :

J mol⁻¹ K⁻¹

	S_{CuO}	S_{Cu_2O}	S_{O_2}
298K	42.64	93.10	205.04

For the dissociation reaction at 298K and 1 atm,

$$\Delta H° = 140,120 \, J$$

$$\Delta G° = 107,150 \, J$$

1.4.2. Absolute Entropies

The equation known as the third law of thermodynamics was originally suggested by W. Nernst :

$$\lim_{T \to 0} \Delta S = 0$$

Consider the phase transformation of metallic tin.

$$Sn(s, \text{grey}) = Sn(s, \text{white})$$

From the Third law,

$$S_{sn(s,\ white)} - S_{sn(s,\ grey)} = 0 \quad \text{at } T = 0 \text{ K}$$

or

$$S_{sn(s,\ white)} = S_{sn(s,\ grey)}$$

However, this does not necessarily mean that

$$S_{sn(s,\ white)} = S_{sn(s,\ grey)} = 0$$

Max Planck offered a different version of the Third law :

$$\boxed{\lim_{T \to 0} S = 0}$$

Planck's hypothesis is based on the statistical thermodynamics in that the entropy is related to the number of possible energy states for a given energy or thermodynamic probability W:

$$\boxed{S = k \ln W}$$

where k is Boltzmann's constant.

At absolute zero there is only one way in which the energy can be distributed in a system; i.e., atoms or molecules and electrons are all in the lowest available quantum states and therefore

$$W = 1$$

Thus

$$S = 0 \quad \text{at } T = 0.$$

Planck's statement of the Third law suggests that *a scale for the absolute value of entropy can be set up* :
1) Set the entropy of a substance equal to zero at $T = 0$ K.
2) Determine the entropy increase from 0 K to the T of interest.
3) Set the result equal to the absolute entropy of the substance at that temperature.

Recall that the entropy change under isobaric conditions is given by

$$\left(\frac{\partial S}{\partial T}\right)_P = C_P dT \quad \text{or} \quad dS = \frac{C_P}{T} dT$$

Integration yields

$$\boxed{S_T - S_0 = \int_0^T \frac{C_P}{T} dT} \quad \text{Since } S_0 = 0, \quad \boxed{S_T = \int_0^T \frac{C_P}{T} dT}$$

Example 1

At low temperatures, especially near absolute zero, data on heat capacities are lacking for many substances. This lack of data is overcome by making extrapolations to lower temperatures. In this regard the following relationships for heat capacity C_P have proved useful at low temperatures:

$C_P = aT^3$: most nonmagnetic, nonmetallic crystals
$C_P = bT^2$: layer lattice crystals, like graphite and boron nitride, and surface heat capacity
$C_P = \gamma T + aT^3$: metals
$C_P = jT^{3/2} + aT^3$: ferromagnetic crystals below the magnetic transition temperature
$C_P = mT^3$: antiferromagnetic crystals below the magnetic transition temperature

For metals γ can be neglected at low temperatures (but not at $T < 1K$). Express entropy for metals at low temperatures as a function of heat capacity.

$$S_T = \int_0^T \frac{C_P}{T} dT = \int_0^T (\gamma + aT^2)dT \quad \xrightarrow{\gamma = 0 \text{ at low } T} \quad S_T = \frac{aT^3}{3} = \frac{C_P}{3}$$

1.5. Calculation of Enthalpies and Free Energies

1.5.1. Standard States

Absolute values of many thermodynamic properties cannot be obtained. This difficulty is overcome by choosing a *reference or standard state* so that properties can be given in terms of the difference between the state of interest and the reference or standard state.

Hypothetical axis for a thermodynamic property.

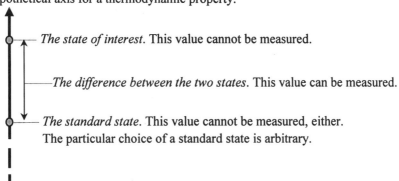

— *The state of interest.* This value cannot be measured.

—*The difference between the two states.* This value can be measured.

— *The standard state.* This value cannot be measured, either. The particular choice of a standard state is arbitrary.

The standard states for solids, liquids and gases which are most commonly used from the point of view of convenience are,

Solid	Liquid	Gas
The most stable, pure substance at 1 atm pressure and the temperature specified.	The most stable, pure substance at 1 atm pressure and the temperature specified.	Ideal gas at 1 atm pressure and the temperature specified.

Example 1

Data on the enthalpy change of titanium (Ti) at 298K is given as follows:

	ΔH
α-Ti	0
β-Ti	3,350 Jmol^{-1}

Which is incorrectly stated in the following?
1) α-Ti is the stable form at 298K.
2) α-Ti is the standard state at 298K.
3) The enthalpy of β-Ti is 3,350 Jmol^{-1}.
4) The enthalpy difference between α- and β-Ti, i.e., $H_{\beta\text{-}Ti} - H_{\alpha\text{-}Ti}$, is 3,350 J mol^{-1}.

The incorrectly stated one is 3). The value of 3,350 Jmol^{-1} merely indicates that the enthalpy of β-Ti is larger than that of α-Ti by 3,350 Jmol^{-1}.

Example 2

Discuss the validity of the following statements :
1. Unless otherwise specified, the standard state of an element *i* customarily chosen to be at a pressure of 1atm and in the most stable structure of that element at the temperature at which it is investigated.
2. However it is possible to choose as a standard state one that does not correspond to the most stable form of the species under consideration.
3. The standard state may also correspond to a virtual state, one that cannot be physically obtained but that can be theoretically defined and for which properties of interest can be calculated.

---

1) This statement describes the general definition of the standard state.
2) This statement is also true. For example, it may be convenient to choose as the standard state of H_2O at 298K that of the gas instead of the liquid, or one may choose at 298K the fcc structure of iron (austenite) rather than the bcc one (ferrite).
3) This statement is also correct. Further discussion in detail is given in Section 3.1.5.

1.5.2. Heat of Formation

We shall recall that the enthalpy change for a process (e.g., a chemical reaction), ΔH, is equal to the value of the heat absorbed or evolved when the process (e.g., reaction) takes place at a constant pressure :

$$\Delta H = q_P$$

We shall also recall that it is not possible to measure the absolute value of a thermodynamic property such as the enthalpy of a substance.
Nevertheless, let's consider a hypothetical system in that it is assumed we know absolute enthalpies of substances at constant temperature and pressure, say,

Species	A	B_2	AB	AB_2
H, Jmol^{-1}	20	30	30	40

Then the enthalpy change for the reaction

$A + B_2 = AB_2$ $\quad \Delta H_1 = H_{AB_2} - (H_A + H_{B_2}) = 40 - (20 + 30) = -10\, J\,mol^{-1}$

Similarly,

$A + 1/2 B_2 = AB$ $\quad \Delta H_1 = H_{AB} - (H_A + 1/2 H_{B_2}) = 30 - (20 + 30/2) = -5\, J\,mol^{-1}$

Those enthalpy values listed in the table are in fact not possible to measure. However, the enthalpy changes of the reactions discussed above can be obtained by measuring heats evolved or absorbed from the reactions. As we are more interested in *enthalpy changes* rather than absolute enthalpies, a new term called *enthalpy of formation* (ΔH_f) or *heat of formation* is introduced.
The enthalpy of formation is defined as the enthalpy change for the reaction in which one mole of the substance is formed for the elements at the temperature of interest.

Species	A	B_2	AB	AB_2
ΔH_f, Jmol^{-1}	0	0	-5	-10

These are zero because they themselves are elements.

If all elements and species are in their standard states, then we use the symbol ΔH_f^o and call it *standard enthalpy of formation*.

Let's check the validity of the concept of the enthalpy of formation by using an example. Consider the following reaction :

$$AB + 1/2\ B_2 = AB_2 \qquad \Delta H_3$$

By using absolute values of H, $\quad \Delta H_3 = 40 - (30 + 30/2) = -5\ \text{Jmol}^{-1}$
By using enthalpy of formation, $\quad \Delta H_3 = -10 - (-5 + 0) = -5\ \text{Jmol}^{-1}$

Note that assigning a value of zero to ΔH_f° for each element in its most stable form at the standard state does not affect our calculations in any way.

Thus, the enthalpies of formation (ΔH_f) or the standard enthalpies of formation (ΔH_f°) which are experimentally measurable are of great practical value and give an easy way of determining the enthalpy change accompanying a process or a reaction.

Example 1

Standard sources of thermodynamic data list the heat or enthalpy of formation at a standard or reference temperature. Most commonly room temperature (298K or 298.15K to be more precise) is chosen as the reference temperature. A few examples are listed in the table given below:

298K, kJmol^{-1}

	Ca(s)*	H$_2$(g)*	H(g)	Sn(s,w)*	Sn(s,g)	CaO(s)	CaCO$_3$(s)
ΔH_f°	0	0	0	0	2.51	-643.3	-1,207.1

"*" represents the standard state.

Find an element or species the standard enthalpy of formation (ΔH_f°) of which is incorrectly specified. Is the following reaction exothermic or endothermic?

$$\text{Sn}(s,\text{white}) \rightarrow \text{Sn}(s,\text{grey})$$

---

1) As H$_2$(g) is specified as the standard state, ΔH_f° of H(g) is not zero. It is experimentally found to be 218kJ mol^{-1}.
2) Sn(s,white) $\rightarrow$ Sn(s,grey) : $\Delta H = 2.51 - 0 = 2.51 > 0$: endothermic reaction.

Example 2

Are the following statements all true?
1) The standard enthalpy of formation is defined as the heat change that results when one mole of a compound is formed from its elements at a pressure of 1 atm.
2) Although the standard state does not specify a temperature, it is customary that we always use ΔH_f° values measured at 25°C (298.15K to be exact).

Both statements are correct. Once ΔH_f° is known at 25°C, then ΔH_f at other temperatures can be calculated using information on the heat capacity. Refer to Section 1.5.3 for further discussion.

1.5.3. Heat of Reaction

Consider enthalpy changes that accompany chemical reactions. The *enthalpy of reaction* or *heat of reaction* is defined as *the difference between the enthalpies of the products and the enthalpies of the reactants* :

$$\Delta H = \Sigma H_{products} - \Sigma H_{reactants}$$

Consider an example of the combustion of methane :

$$CH_4 + 2O_2 = CO_2 + 2H_2O \qquad \Delta H_r$$

```
┌──────────────┐
│  CH₄ + 2O₂   │ ─────────→  ΣH(reactants) = H_CH₄ + 2H_O₂
└──────┬───────┘
       │ Heat given off
       │ at constant P and T   →  ΔH_r
       ▼
┌──────────────┐
│  CO₂ + 2H₂O  │ ─────────→  ΣH(products) = H_CO₂ + 2H_H₂O
└──────────────┘
```

Thus

$$\Delta H_r = (H_{CO_2} + 2H_{H_2O}) - (H_{CH_4} + 2H_{O_2})$$

→ However we are not able to measure these values of absolute enthalpies.

The enthalpy of formation, which we have discussed in the previous section, offers an easy way to overcome this difficulty. We now introduce *Hess's Law*. Recall that enthalpy is a state function, and hence the enthalpy change depends on only the initial and final states. Hess's law is basically the same as stated above, but expressed in a different way : Hess's law states that the enthalpy change for a chemical reaction is the same whether it takes place in one or several stages.

Consider the combustion of methane again.

$$CH_4 + 2O_2 = CO_2 + 2H_2O \qquad \Delta H_r$$

$$CH_4 = C + 2H_2 \qquad : -\Delta H_{f,CH_4}$$

$$2O_2 = 2O_2 \qquad : 2\Delta H_{f,O_2} = 0$$

$$C + O_2 = CO_2 \qquad : \Delta H_{f,CO_2}$$

$$2H_2 + O_2 = 2H_2O \qquad : 2\Delta H_{f,H_2O}$$

+) ―――――――――――――――――――――――――――

$$CH_4 + 2O_2 = CO_2 + 2H_2O \qquad : -\Delta H_{f,CH_4} + \Delta H_{f,CO_2} + 2\Delta H_{f,H_2O}$$

Therefore

$$\Delta H_r = -\Delta H_{f,CH_4} + \Delta H_{f,CO_2} + 2\Delta H_{f,H_2O}$$

or

$$\Delta H_r = (\Delta H_{f,CO_2} + 2\Delta H_{f,H_2O}) - (\Delta H_{f,CH_4})$$

<u>Weighted sum of the enthalpies of formation of the products</u>   <u>Weighted sum of the enthalpies of formation of the reactants</u>

Now it is in order to generalise our discussion. Consider the following reaction :

$$aA + bB = cC + dD \qquad \Delta H_r$$

The heat of reaction is given by

$$\Delta H_r = (c\Delta H_{f,C} + d\Delta H_{f,D}) - (a\Delta H_{f,A} + b\Delta H_{f,B})$$

This equation applies to a system undergoing a chemical reaction at *constant pressure* and *temperature*.
- If $\Delta H_r > 0$, the reaction takes place with an absorption of heat from surroundings (*endothermic*).
- If $\Delta H_r < 0$, the reaction takes place with an evolution of heat to the surroundings (*exothermic*).

If temperature is not constant, but changes, recall that

$$\left(\frac{\partial H}{\partial T}\right)_P = C_P \quad \text{or} \quad dH = C_P dT \quad \text{at constant } P.$$

For a system undergoing a temperature change from T_1 to T_2,

$$\Delta H = \int_{T_1}^{T_2} C_P dT$$

Consider a general reaction which occurs at a constant pressure :

$$aA + bB = cC + dD$$

Enthalpy change of reactants (a moles of A and b moles of B) undergoing temperature change from T_1 to T_2. $\Delta H_{react} = H_{T_2,react} - H_{T_1,react}$ $= \int_{T_1}^{T_2} (aC_{P,A} + bC_{P,B}) dT$	Enthalpy change of products (c moles of C and d moles of D) undergoing temperature change from T_1 to T_2. $\Delta H_{prod} = H_{T_2,prod} - H_{T_1,prod}$ $= \int_{T_1}^{T_2} (cC_{P,C} + dC_{P,D}) dT$

But the enthalpy change of reaction at T_2 is

$$\Delta H_{r,T_2} = H_{T_2,prod} - H_{T_2,react}$$

$$H_{T_2,prod} = H_{T_1,prod} + \int_{T_1}^{T_2} (cC_{P,C} + dC_{P,D})dT$$

$$H_{T_2,react} = H_{T_1,react} + \int_{T_1}^{T_2} (aC_{P,A} + bC_{P,B})dT$$

Thus

$$\boxed{\Delta H_{r,T_2} = \Delta H_{r,T_1} + \int_{T_1}^{T_2} \Delta C_P dT}$$

where $\Delta H_{r,T_1}$ = heat of reaction at T_1, and

$$\Delta C_P = (cC_{P,C} + dC_{P,D}) - (aC_{P,A} + bC_{P,B})$$

This equation is known as *Kirchhoff's law* in integral form. It enables us to calculate the heat of reaction at different temperatures by knowing the heat of reaction at one temperature, say, 298K, and heat capacities of reactants and products.
In a differential form,

$$\boxed{\left(\frac{\partial \Delta H}{\partial T}\right)_P = \Delta C_P}$$

Example 1

Consider a general reaction $aA + bB = cC + dD$.
The following figure shows the change of enthalpy of reactants and the products with temperature. Answer to each of the following questions :

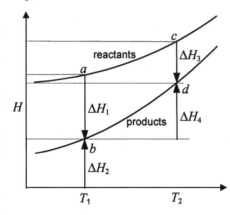

1) Which point represents the sum of enthalpies of reactants at T_1?
2) Which point represents the sum of enthalpies of the products at T_2?
3) Which one represents the enthalpy change of the products with change in temperature from T_1 to T_2?
4) Which one is the heat of reaction at T_2?
5) Is the reaction endothermic or exothermic at T_2?

---

1) a 2) d 3) $\Delta H_4 (= d - b)$ 4) $\Delta H_3 (= d - c)$ 5) Exothermic ($\Delta H_3 < 0$)

Example 2

The diagram shows the variation with temperature of enthalpy for a general oxidation reaction :

$$M + 1/2 O_2 = MO.$$

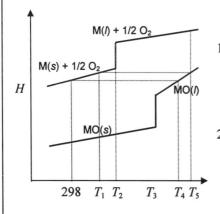

1) One mole of M(s) reacts to completion with 1/2 mole of $O_2(g)$ at 298K under an adiabatic condition. Which point represents the temperature of the product MO?
2) Under the adiabatic condition, it is desired to raise the temperature of the product MO to T_5. To what temperature would the reactions need to be preheated?
3) Which part of the diagram represents the heat of fusion of M?

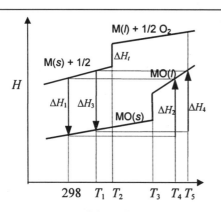

1) Heat liberated by the reaction = ΔH_1
This heat is used up in raising the temperature of the product MO.
When the temperature of MO is raised to T_4, the amount of heat abosorbed by MO is ΔH_2.

$$\Delta H_1 + \Delta H_2 = 0.$$

∴ Product temperature = T_4.

2) Amount of heat liberated by the reaction = Amount of heat absorbed by the product (adiabatic)
If the reaction takes place at T_1, the amount of heat liberated is ΔH_3. This heat is enough to raise the temperature of the product MO from T_1 to T_5 (ΔH_4). Therefore the reactants need to be preheated to T_1.
3) Fusion is an isothermal phase transformation and hence the enthalpy change of M increases without accompanying a temperature change. ΔH_t represents the heat of fusion of M.

Exercises

1. The enthalpy change associated with freeze of water at 273K is -6.0kJmol⁻¹. The heat capacity (C_p) for water is 75.3Jmol⁻¹K⁻¹ and for ice 37.6Jmol⁻¹K⁻¹. Calculate the

enthalpy change when water freezes at 263K.

2. The enthalpy changes at 298K and 1atm for the hydrogenation and for the combustion of propane are given below :

    $C_3H_6(g) + H_2(g) = C_3H_8(g)$ $\Delta H_1 = -124 \text{kJmol}^{-1}$
    $C_3H_8(g) + 5O_2(g) = 3CO_2(g) + 4H_2O(l)$ $\Delta H_2 = -2,220 \text{kJmol}^{-1}$

    In addition the enthalpy change at 298K and 1atm of the following reaction is known :

    $H_2(g) + 1/2O_2(g) = H_2O(l)$ $\Delta H_3 = -286 \text{kJmol}^{-1}$

    Calculate the heat liberated by the combustion of one mole of propane at 298K and 1atm.

3. The extraction of zinc by carbothermic reduction of zinc oxide sinter at 1,100°C can be represented by the reaction

    $$ZnO(s) + C(s) = Zn(g) + CO(g)$$

    Calculate the heat of reaction at 1,100°C. the following data are given :

	$\Delta H°_{f,298}$ (kJmol^{-1})	T_t (K)	ΔH_t (kJmol^{-1})	C_P (Jmol^{-1}K^{-1})
ZnO(s)	-348.1			$49.0 + 5.10 \times 10^{-3}T - 9.12 \times 10^{5}T^{-2}$
CO(g)	-110.5			$28.4 + 4.10 \times 10^{-3}T - 0.46 \times 10^{5}T^{-2}$
C(s)	0			$17.2 + 4.27 \times 10^{-3}T - 8.8 \times 10^{5}T^{-2}$
Zn(s)	0	693 ($s \to l$)	7.28	$22.4 + 10.0 \times 10^{-3}T$
Zn(l)	-	1,180 ($l \to g$)	114.2	31.4
Zn(g)	-			20.8

1.5.4. Adiabatic Flame Temperature

Exothermic chemical reactions can be used as energy sources. The more heat evolved per unit mass of fuel, the greater the utility of the fuel as an energy source.
The heat released by the combustion raises the temperature of the combustion products. If there is no heat loss to the surroundings and all heat goes into raising the temperature of the products, then the flame temperature will be the highest. This flame temperature is called the *adiabatic flame temperature*.

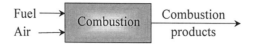

For a steady-flow adiabatic combustion system, the total enthalpy is conserved:

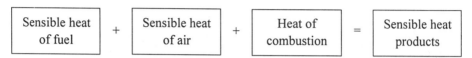

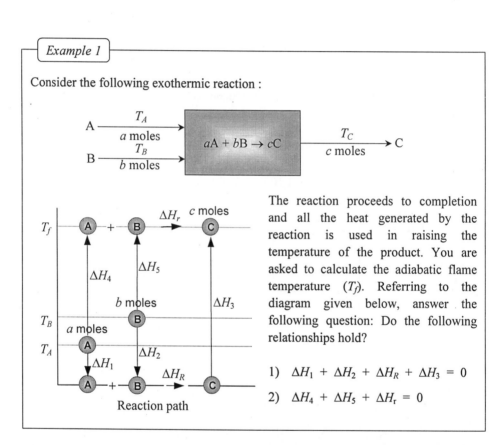

Example 1

Consider the following exothermic reaction:

The reaction proceeds to completion and all the heat generated by the reaction is used in raising the temperature of the product. You are asked to calculate the adiabatic flame temperature (T_f). Referring to the diagram given below, answer the following question: Do the following relationships hold?

1) $\Delta H_1 + \Delta H_2 + \Delta H_R + \Delta H_3 = 0$

2) $\Delta H_4 + \Delta H_5 + \Delta H_r = 0$

1) The equation is correct because the total enthalpy is conserved and the enthalpy is a state function so that the enthalpy change is independent of the path the process takes as long as the initial and final states are kept unchanged. The adiabatic flame temperature T_f can be found by solving the equation for T_f.

2) This equation is also correct. We can construct several different paths between the initial and final states because enthalpy is a state function.

Exercises

1. A fuel gas containing 22% CO, 13% CO_2 and 65% N_2 by volume is combusted with the theoretically required amount of air in a furnace to heat a solid burden. The gases enter the furnace at 250°C and the following data are available :

	$\Delta H°_{298}$ (Jmol^{-1})	C_P (Jmol^{-1}K^{-1})
CO(g)	-110,530	$28.41 + 4.1 \times 10^{-3}T - 0.46 \times 10^5 T^{-2}$
CO_2(g)	-393,510	$44.14 + 9.04 \times 10^{-3}T - 8.54 \times 10^5 T^{-2}$
O_2(g)		$29.96 + 4.18 \times 10^{-3}T - 1.67 \times 10^5 T^{-2}$
N_2(g)		$27.87 + 4.27 10^{-3}T$

Calculate the adiabatic flame temperature.

1.5.5. Gibbs Free Energy Changes

The value of the Gibbs free energy change for a chemical reaction at the temperature specified is given by the difference in Gibbs free energy between the products and the reactants.
For reaction

$$aA + bB = cC + dD \qquad \Delta G_r$$

$$\Delta G_r = (cG_C + dG_D) - (aG_A + bG_B)$$

where ΔG_r is the Gibbs free energy change of the reaction, $(cG_C + dG_D)$ is the Total Gibbs free energy of the products, and $(aG_A + bG_B)$ is the Total Gibbs free energy of the reactants.

If the reactants and products are all in their standard states at the temperature specified,

$$\Delta G_r^o = (cG_C^o + dG_D^o) - (aG_A^o + bG_B^o)$$

As in the case of enthalpies, it is not possible to obtain the absolute values of these standard free energies. However, this difficulty can be overcome by introducing a term called *standard free energy of formation*.

Standard free energy of formation: ΔG_f^o

The standard Gibbs free energy change of reaction for the *formation of a compound from its elements* at their standard states.

As in the case of enthalpy, the standard free energy of formation of each element is put "zero" at all temperatures.

For example,

$$C(s) + 1/2 O_2(g) = CO(g) \qquad \Delta G_{f,CO}^o$$

$$C(s) + O_2(g) = CO_2(g) \qquad \Delta G_{f,CO_2}^o$$

$$CO(g) + 1/2 O_2(g) = CO_2(g) \qquad \Delta G_r^o$$

$$\Delta G_r^o = \Delta G_{f,CO_2}^o - \left(\Delta G_{f,CO}^o + \frac{1}{2} \Delta G_{f,O_2}^o \right)$$

zero because it is an elemental species.

Thus the standard free energy change of reaction can be calculated from the standard free energy of formation of species involved in the reaction.

Reminder

ΔG_r^o is the change of Gibbs free energy which occurs when all the reactants and products are in their standard states. To deal with systems in which there are some species not in their standard states, but in mixtures (solutions), we shall have to consider *partial properties* which will be discussed later.

From the definition of Gibbs free energy,

$$G = H - TS$$

Thus, for a chemical reaction taking place under constant temperature T,

$$\Delta G_r = \Delta H_r - T\Delta S_r$$

If all the reactants and the products are in their respective standard states at constant temperature T,

$$\Delta G_r^o = \Delta H_r^o - T\Delta S_r^o$$

Example 1

Consider the following reaction of formation of species AB_2:

$$A + 2B = AB_2 \qquad \Delta G^o_{f,AB_2} \qquad T = \text{constant}$$

The following statements are all true. Discuss the application of the statements using the example of the formation of $TiSi_2(s)$.

1) $\Delta G^o_{f,AB_2}$ is the difference between the free energy of 1 mole of AB_2 and the sum of the free energies of 1 mole of A and 2 moles of B, all in their standard states at temperature T.
2) $\Delta G^o_{f,AB_2}$ is the standard free energy of formation of compound AB_2 measured with the free energy scale established by setting

$$\Delta G^o_{f,A} = 0 \quad \text{and} \quad \Delta G^o_{f,B} = 0.$$

3) The state of each element (i.e., A or B) for which the above relationships are set, that is, the standard state of each element, is arbitrarily chosen.
4) The form of element A or B for which the free energy of formation is taken zero must be the same form for which $\Delta H^o_f = 0$ is taken.

---

1) $Ti + 2Si = TiSi_2 \qquad \Delta G^o_{f,TiSi_2} = G^o_{TiSi_2} - (G^o_{Ti} + 2G^o_{Si})$
2) $\Delta G^o_{f,TiSi_2}$ is the standard free energy of formation of $TiSi_2$ measured with the basis of $\Delta G^o_{f,Ti} = 0$ and $\Delta G^o_{f,Si} = 0$.
3) The standard state of a species can be arbitrarily chosen
4) As G is related with H by $G = H - TS$, the standard state should be consistent.

Example 2

Consider the following reaction:

$$AB + B = AB_2 \qquad \Delta G^o_r \qquad T = \text{constant}$$

Check the validity of each of the following statements:

1) ΔG^o_r is the difference between the free energy of 1 mole of AB_2 and the sum of the free energies of 1 mole of AB and 1 mole of B, all in their standard states at temperature T.
2) The choice of a standard state is arbitrary. If we choose the most stable and pure form of AB at temperature T as its standard state, however, we must choose the most stable and pure forms of all other species (i.e., B and AB_2) as their respective standard states.

1) The standard free energy change of a reaction (ΔG_r^o) is the net free energy change resulted from the reaction. The statement is true.
2) This statement is not true. Each species can take any state as its standard state, but one must be consistent in choosing the standard state of a species throughout the computations

Exercises

1. Calculate the standard Gibbs free energy change of the following reaction at 1,000K and 1 atm :

$$CO(g) + 1/2\ O_2(g) = CO_2(g) \qquad \Delta G_r^o$$

The following data are available :

$$\Delta G_{f,CO}^o = -112,880 - 86.51T,\ \text{Jmol}^{-1}$$
$$\Delta G_{f,CO_2}^o = -394,760 - 0.836T,\ \text{Jmol}^{-1}$$

2. Compute ΔG_r^o for the reaction

$$CH_4(g) + 2O_2(g) = CO_2(g) + 2H_2O(g)$$

at 298K. The following data are given :

	$\Delta H_{f,298}^o$ Jmol^{-1}	$S_{f,298}^o$ Jmol^{-1}K^{-1}
$CH_4(g)$	-74,850	186.2
$CO_2(g)$	-393,510	213.7
$H_2O(g)$	-241,810	188.7
$O_2(g)$	0	205.0

CHAPTER 2

SOLUTIONS

2.1. Behaviour of Gases

2.1.1. Ideal Gases

An experimentally found relationship is that,

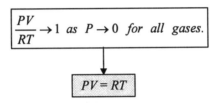

This equation relates the state variables of the system. This equation is called the *ideal gas equation* or *perfect gas equation*. A gas, which obeys this relationship over a range of states of interest, is said to *behave ideally* in the range. A gas, which obeys this relationship in all possible states, is called an *ideal gas* or a *perfect gas*.

Recall the following equation :

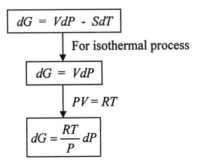

Integration from the standard state ($P = 1$ atm) to the state of interest at constant temperature yields

$$G = G^\circ + RT \ln P$$

Where G° is molar free energy of an ideal gas in its standard state and G is molar free energy of an ideal gas in the state of interest.

We now consider mixtures of ideal gases.

Chemical Thermodynamics for Metals and Materials

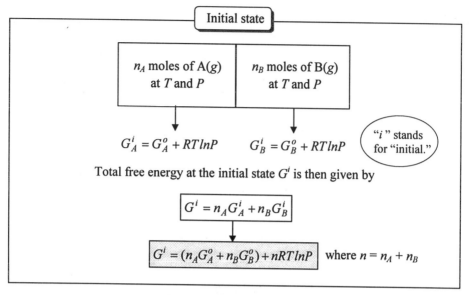

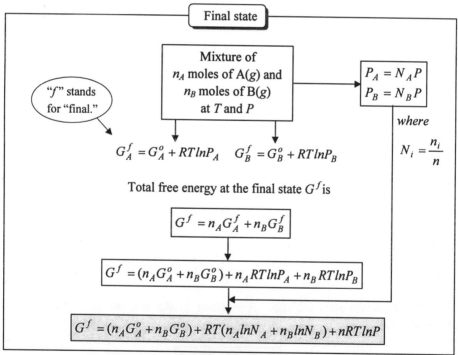

Thus, the change in Gibbs free energy on mixing, G^m

$$G^m = G^f - G^i$$

From the equations developed above

$$G^m = RT(n_A \ln N_A + n_B \ln N_B)$$

For one mole of the mixture, $G^M = \dfrac{G^m}{n}$,

$$G^M = RT(N_A \ln N_A + N_B \ln N_B)$$

Summary

$G^M = RT(N_A \ln N_A + N_B \ln N_B)$ Gibbs free energy change for the formation of one mole of mixture or solution.

$G_A = G_A^o + RT \ln P_A$ Gibbs free energy of one mole of A after mixing or in solution

$G_B = G_B^o + RT \ln P_B$ Gibbs free energy of one mole of B after mixing or in solution

G_A and G_B (later we will be using symbols $\overline{G}_i$) are called *partial molar free energies* of A and B, respectively. These are also called *chemical potentials* (μ_i). G^M is called *relative molar free energy* of the solution. A fuller discussion on these properties is given in Section 2.2.2.

Note that $G^M < 0$ because $N_i < 0$. Therefore mixing is a spontaneous process.

For S^M and H^M,

$$S = -\left(\dfrac{\partial G}{\partial T}\right)_P \longrightarrow \boxed{S^M = -R(N_A \ln N_A + N_B \ln N_B)}$$

$G = H - TS \longrightarrow \boxed{H^M = 0}$

↳ For ideal gases, no heat is evolved, or absorbed in the mixing process.

Exercises

1. n_A moles of gas A and n_B moles of gas B are mixed at constant temperature and pressure. Calculate the ratio of A and B which minimises the free energy of the mixture. Assume that both A and B are ideal gases.

2. A 1m³ cylinder contains $H_2(g)$ at 298K and 1atm, and is connected to another cylinder which contains 3m³ $O_2(g)$ at 298K and 0.8atm. When the valve is opened, the gases diffuse into each other and form a homogeneous mixture under isothermal conditions. Calculate the free energy change of mixing, G^m, for the process. Assume the gases behave ideally.

2.1.2. Fugacities and Real Gases

In the previous section we have developed the following equation for an ideal gas:

$$G = G^o + RT\ln P$$

If a gas deviates from ideality, this equation ceases to apply. However, it is desired to preserve this simple form of expression as much as possible for non-ideal, real gases. The above equation shows that the free energy G is a linear function of the logarithm of the pressure of an ideal gas. Now, let's introduce a function which when used in place of the real pressure ensures linearity between G and the logarithm of this function in any state of any gas. This function is called *fugacity* (f), a sort of corrected pressure.

$$G = G^o + RT\ln f$$

How do we evaluate the fugacity f?

$$\boxed{dG = VdP - SdT}$$

constant T

$$\boxed{dG = VdP}$$

We now express V as a function of P:
- Volume of real gas : V_{real}
- If the gas were ideal, the volume would be : $V_{ideal} = RT/P$
- The difference (α) is : $\alpha = V_{ideal} - V_{real} = RT/P - V_{real}$
- Rearrangement yields : $V = RT/P - \alpha$ where $V = V_{real}$

$$\boxed{dG = \left(\frac{RT}{P} - \alpha\right)dP}$$

$dG = RT\, d\ln f$

$$\boxed{RT\, d\ln f = \left(\frac{RT}{P} - \alpha\right)dP}$$

Integration from $P = 0$ to a state of interest,

$$\boxed{\int_{P=0}^{P=P} RT\, d\ln f = \int_{P=0}^{P=P} \left(\frac{RT}{P} - \alpha\right)dP}$$

Rearrangement yields

$$\left[RT\ln\left(\frac{f}{P}\right)\right]_{P=0}^{P=P} = -\int_{P=0}^{P=P} \alpha\, dP$$

Knowing that $f/P \to 1$ as $P \to 0$.

$$RT\ln\left(\frac{f}{P}\right) = -\int_{P=0}^{P=P} \alpha\, dP$$

This equation enables us to evaluate the fugacity at any pressure and temperature, provided that data on PVT for the gas of interest are available.

Graphical Method
1. Plot the deviation (α) from ideality of the gas against P.
2. Evaluate the area between the integration limit.

Analytical Method
1. Express α as a function of P.
2. Evaluate the integral analytically.

We now discuss more on the analytical method. Rearrangement of the above equation yields

$$\ln\left(\frac{f}{P}\right) = -\int_{P=0}^{P=P} \left(\frac{\alpha}{RT}\right) dP$$

$\alpha = RT/P - V$

We introduce a term called the *compressibility factor Z* defined as

$$Z = \frac{PV}{RT}$$

$$\ln\left(\frac{f}{P}\right) = \int_{P=0}^{P=P} \frac{Z-1}{P} dP$$

Z is 1 for ideal gases, but for real gases, it is a function of the state of the system, e.g., $Z = f(P,T)$. Some equations of state for non-ideal gases are given below:

$$\frac{PV}{RT} = 1 + B_2 P + B_3 P^2 + \cdots$$

$$\frac{PV}{RT} = 1 + \frac{C_2}{V} + \frac{C_3}{V^2} + \cdots$$

where B_i's and C_i's are called *virial coefficients* and depend only on the temperature.

At low pressure or densities the first two terms in the state equations are sufficient to represent the state:

$$\frac{PV}{RT} = 1 + B_2 P \qquad \frac{PV}{RT} = 1 + \frac{C_2}{V}$$

Thus it is now possible to evaluate analytically the integral in the previous equation.

Example 1

A real gas obeys the following equation of state:

$$PV = RT + BP$$

where B is independent of pressure and is a function of temperature only. Choose an incorrect one from the following relationships:

$$\frac{f}{P} = \exp\left(\frac{BP}{RT}\right) \qquad \frac{f}{P} = 1 + \frac{BP}{RT} \qquad \frac{f}{P} = \ln\left(\frac{BP}{RT}\right) \qquad \frac{f}{P} = Z$$

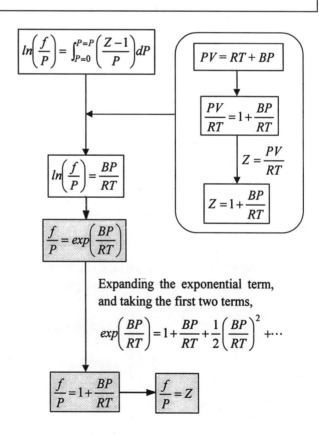

Example 2

The most powerful equation of state that describes the behaviour of real gases rather well is the *van der Waals equation* :

$$\left(P+\frac{a}{V^2}\right)(V-b)=RT$$

where a and b are constants which are characteristic of each gas. This equation was derived by taking into considerations the properties of real gases :

Difference in physical properties between ideal gases and real gases

	Ideal gases	Real gases
Particle volume	Volumeless	A finite volume.
Particle interaction	No interaction	Interaction

In the above equation, which constant is related to the fact that the particles of a real gas occupy a finite volume? Which constant in the equation is related to the fact that interactions occur among the particles of a real gas?

The constant b is related to the correction for the finite volume of the particles in a real gas, and the constant a is related to the correction for particle-particle interactions.

Exercises

1. The virial equation for hydrogen gas at 298K in the pressure range 0 to 1,000atm is given below :

$$PV = RT(1 + 6.4\times10^{-4}P)$$

   1) Calculate the fugacity of hydrogen at 100atm and 298K.
   2) Calculate the free energy change associated with the compression of 1 mole of hydrogen at 298K from 1atm to 100atm.

2.2 Thermodynamic Functions of Mixing

2.2.1. Activities and Chemical Potentials

We have proved in Section 2.1.2. that fugacity, as a sort of modified pressure, plays a valuable role in dealing with non-ideal, real gases. We extend the concept of fugacity to condensed phases, i.e., solids and liquids. Let's begin our discussion with the vaporisation process of substance A in a condensed phase.

$$A(c) = A(g) \qquad \Delta G_A$$

Where "c" is for a condensed phase, and ΔG_A is the molar Gibbs free energy change for the vaporisation process.

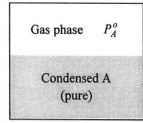

At equilibrium the pressure of A in the gas phase is the saturation vapour pressure of A at temperature T, and hence

$$\boxed{\Delta G_A = G^*_{A(g)} - G^o_{A(c)} = 0}$$

Molar free energy of gaseous A at $P = P^o_A$ Molar free energy of condensed A at the pure state.

Next, consider the vaporisation of A not in the pure state, but in the state of solution.

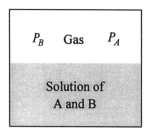

The vapor pressure of A in the gas phase will be P_A at temperature T, and

$$\boxed{\Delta G_A = G_{A(c)} - G_{A(g)} = 0}$$

Molar free energy of A in the solution   Molar free energy of A in the gas phase

Now we consider the free energy change associated with the change in state of A from pure condensed state to the state of liquid or solid solution.

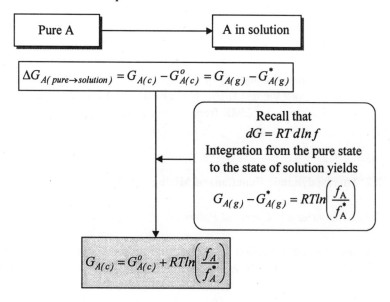

Note that this equation relates the free energy change in the condensed phase to the fugacities of the vapour phase. For substances which have rather low vapour pressures,

$$f \cong P$$

$$G_{A(c)} = G^o_{A(c)} + RT \ln\left(\frac{P_A}{P^o_A}\right)$$

Now we define the fugacity of a substance in condensed phase :

| Fugacity of a substance in the condensed phase, i.e., solid of liquid. | = | Fugacity of the vapour that is in equilibrium with the substance. |

We introduce a function called *activity*, a, which is defined as

$$a_i = \frac{f_i}{f_i^o}$$ where i is the i-th component in solution.

In most cases of chemical reaction systems, particularly in metallurgical systems, vapour pressures are not high so that
$$f = P$$

$$a_i = \frac{P_i}{P_i^o}$$

Using the new function of activity, we have

$$G_{A(c)} = G^o_{A(c)} + RT \ln\left(\frac{P_A}{P^o_A}\right) \longrightarrow G_{A(c)} = G^o_{A(c)} + RT \ln a_A$$

In the equations, the term $G_{A(c)}$ is the molar free energy of A in the solution with the concentration which exerts the vapour pressure of P_A. This term is called *partial molar free energy* of A and denoted using the symbol $\overline{G}_A$. It is also called *chemical potential* of A and expressed by symbol μ_A.

Thus for i-th component in a solution,

$$\overline{G}_i = \mu_i = G_i^o + RT \ln a_i$$

This equation is true for any solutions: gas, liquid or solid solutions; the only approximation made is the use of pressures rather than fugacities.

For a gaseous phase, P_i^o is the pressure of species i at the standard state, i.e., $P_i^o = 1$ atm. Thus

$$a_i = P_i$$

Note that the activity of a gaseous species is numerically the same as the pressure in atm unit of the species.

We next examine the change in vapour pressure with the change in composition in a solution.

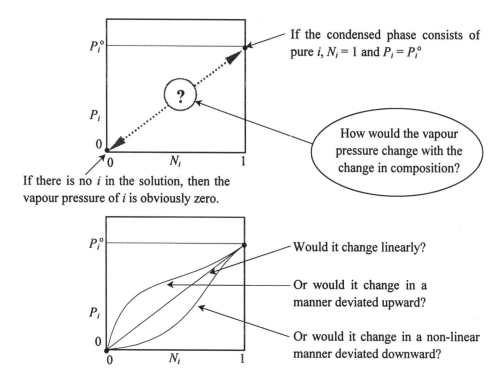

The answer is that it depends on how species i interacts with other species in the solution. Atoms (or molecules) in the solution interact with their neighbouring atoms (or molecules). Consider an i-j binary solution. If i-i, i-j and j-j interactions are all identical, the vapour pressure will be linearly proportional to the concentration. If the i-j attraction is weaker than the i-i interaction, i will become freer by having j as some of its neighbours, and hence more active in the solution and easier to vaporise (upward deviation). If the i-j attraction is stronger than the i-i attraction, on the other hand, i will become more bound by having j as some of its neighbours, and hence less active and more difficult to vaporise (downward deviation).

From the definition of activity, we can convert P_i-N_i relationships to a_i-N_i relationships:

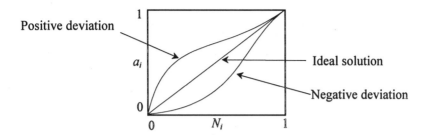

Activity may be regarded as *active* or *effective concentration*, and can be related to the actual concentration by the equation

$$a_i = \gamma_i N_i$$

where γ_i is called *activity coefficient*.

The value of the activity coefficient is the barometer of the extent of deviation from ideal behaviour and the determination of γ is of prime importance in chemical thermodynamics. The variation of the activity coefficient with temperature and composition is generally determined experimentally.

Example 1

Is the following statement true?
"When activity is defined by

$$a_i = \frac{P_i}{P_i^o}$$

P_i^o is the saturation vapor pressure of species i which is in equilibrium with pure i at the temperature of interest *irrespective of* the standard state chosen."

The statement is incorrect. P_i^o is the pressure of species of i at the standard state chosen. If i is a component in the gaseous phase, $P_i^o = 1$ atm. If i is a component in the liquid or solid solution, P_i^o is the vapour pressure which is in equilibrium with species i which is at the standard state. Therefore only if the pure state is chosen at the standard state for the species i, the statement given is true.

Example 2

The activity of a species is unity when it is at the standard state. This is true, even though the choice of the standard state is arbitrary. Is this statement true?

$$a_i = \frac{f_i}{f_i^o} \quad \text{or} \quad a_i = \frac{P_i}{P_i^o}$$

$f_i = f_i^o$ or $P_i = P_i^o$ at the standard state.

$$a_i = 1$$

irrespective of
the choice of the standard state.

Example 3

Is the following statement true?
Although the partial molar Gibbs free energy ($\overline{G}_i$) or the chemical potential (μ_i) defined by the equation given below pertains to the individual components of the system, it is a property of the system as a whole. The value of the partial molar free energy depends not only the nature of the particular substance in question but also on the nature and relative amounts of the other components present as well.

$$\overline{G}_i = \mu_i = G_i^o + RT\ln a_i$$

In the equation, G_i^o is independent of the system, but dependent only on the standard state chosen for the component i. However the activity a_i is dependent on the system, because its value is affected by the nature of interaction with the other components in the system.

Example 4

Is the following statement true?
The value of the partial molar free energy or the chemical potential of a component in the solution is independent of the choice of the standard state for the component.

The statement is true.

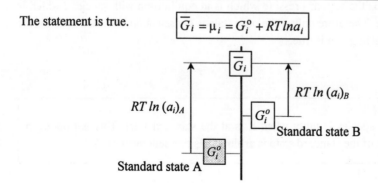

Exercises

1. The vapour pressure of pure solid silver and solid silver-palladium alloys are given in the following :
   For silver,

   $$\log P = \frac{-13{,}700}{T} + 8.73 \qquad \text{(torr)}$$

   For the solid silver-palladium alloy at $N_{Ag} = 0.8$,

   $$\log P = \frac{-13{,}800}{T} + 8.65 \qquad \text{(torr)}$$

   1) Calculate the activity of silver in the alloy at 1,150K. Pure solid silver is taken as the standard state for silver in the alloy.

   2) Calculate the activity coefficient of silver in the alloy at 1,150K.

2. In A-B binary solutions at 600K, the vapour pressures of A at different compositions are

   atm

N_A	1.0	0.9	0.6	0.4	0.3	0.2	0.1
$P_A \times 10^4$	4.9	4.2	2.0	0.9	0.6	0.4	0.2

   Assuming the free energy of pure A at temperature 600K is set to zero, calculate the chemical potential of A in the solution of $N_A = 0.6$.

3. The activity coefficient of zinc in liquid zinc-copper alloys in the temperature range 1,070 - 1,300K can be expressed as follows :

   $$RT\ln\gamma_{Zn} = -31{,}600 N_{Cu}^2 \qquad \text{where } R = 8.314 \text{ J mol}^{-1}\text{K}^{-1}.$$

   Calculate the vapour pressure of Zn over a Cu-Zn binary solution of $N_{zn} = 0.3$ at 1,280K. The vapour pressure of pure liquid zinc is given by

   $$\log P = \frac{-6{,}620}{T} - 2.26\log T + 12.34 \qquad \text{(torr)}$$

2.2.2. Partial Properties

We consider mixtures of substances, i.e., solutions, that do not react with each other. Suppose we have substance A :

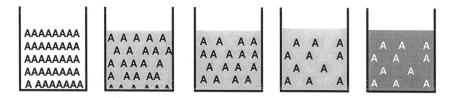

- Would A in a solution behave the same as in the pure state?
- How would A behave in the solution if the concentration is different?
- What if the concentration is the same, but other substances in the solution are different?

In general, thermodynamic properties of the components in a solution vary with composition because the environment of each type of atom or molecule changes as the composition changes. The change in interaction force between neighbouring atoms or molecules with the change in composition results in the variation of the thermodynamic properties of a solution. The thermodynamic properties that components have in a solution are called *partial properties*.

In many processes, we are concerned with mixtures, i.e., gas, liquid or solid solutions, and hence we are concerned with the thermodynamic properties of a component in a solution - *partial molar quantities*:

> Partial molar volume (V)
> Partial molar energy (U)
> Partial molar enthalpy (H)
> Partial molar entropy (S)
> Partial molar free energy (G)

We begin our discussion with volume, since it is perhaps easiest to visualise. The results obtained will then be extended to other quantities. We will also consider a binary-two component-system initially, and then generalise the results for multi-component systems.

The molar volume of a pure substance can be measured with ease. When the same amount of the substance enters a solution, however, it may not take up the same volume. For example,

Water		Ethanol		Water + Ethanol
100 mL	+	100 mL	=	200 mL ?
25°C		25°C		25°C

No. The volume after mixing is not 200mL, but about 190mL.

As seen in this example, the volume of a solution is not, in general, simply the sum of the volumes of the individual components.

Consider a binary solution containing n_A moles of A and n_B moles of B. The volume of the solution is V.

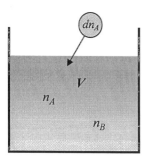

We now add a very small amount of A, dn_A, so small that the addition of this extra amount does not change the concentration of the solution to an appreciable extent. The resulting increase in the volume of the solution at constant temperature and pressure is dV. dV can then be regarded as the volume occupied by dn_A moles of A in the solution of the particular composition. In other words, dn_A moles of A acts in the solution as if it had the volume of dV.

Alternatively we may say that one mole of A in the solution acts as if it has the volume of

$$\frac{dV}{dn_A}$$

This volume is called *partial molar volume* of A and denoted by symbol. $\overline{V}_A$. This partial volume is dependent on the composition of the solution. If the composition changes, this value also changes. As T, P and n_B are kept constant during the process, we may define the partial molar volume as

$$\overline{V}_A = \left(\frac{\partial V}{\partial n_A}\right)_{P,T,n_B}$$

The contribution of A to the total volume is then $n_A \overline{V}_A$ and the contribution of B to the total volume is $n_B \overline{V}_B$. Thus

$$V = \overline{V}_A n_A + \overline{V}_B n_B$$

Volume is a state function with variables of T and P for a closed system, or with variables T, P, n_A, and n_B for an open system.

Closed system	*Open system*
A system which does not allow transfer of matter between the system and the surroundings.	A system which does allow transfer of matter between the system and the surroundings.

As $V = V(T,P,n_A,n_B)$, the complete differential at constant T and P yields

$$dV = \left(\frac{\partial V}{\partial n_A}\right)_{P,T,n_B} dn_A + \left(\frac{\partial V}{\partial n_B}\right)_{P,T,n_A} dn_B$$

$$dV = \overline{V}_A dn_A + \overline{V}_B dn_B$$

Recall that

$$V = \overline{V}_A n_A + \overline{V}_B n_B$$

Differentiation of this equation yields

$$dV = \overline{V}_A dn_A + \overline{V}_B dn_B + n_A d\overline{V}_A + n_B d\overline{V}_B$$

Comparison of the above two differential equations yields

$$n_A d\overline{V}_A + n_B d\overline{V}_B = 0$$

Dividing by the total number of moles, n,

$$N_A d\overline{V}_A + N_B d\overline{V}_B = 0$$

It is also readily seen that

$$V = \overline{V}_A N_A + \overline{V}_B N_B$$

where V is the molar volume of the solution or *integral molar volume*.

Similar equations may be developed for other thermodynamic properties:

	Enthalpy	Entropy	Free energy
Definition of partial properties	$\overline{H}_A = \left(\dfrac{\partial H}{\partial n_A}\right)_{P,T,n_B}$	$\overline{S}_A = \left(\dfrac{\partial S}{\partial n_A}\right)_{P,T,n_B}$	$\overline{G}_A = \left(\dfrac{\partial G}{\partial n_A}\right)_{P,T,n_B}$
Integral molar properties	$H = \overline{H}_A N_A + \overline{H}_B N_B$	$S = \overline{S}_A N_A + \overline{S}_B N_B$	$G = \overline{G}_A N_A + \overline{G}_B N_B$
Relationships of partial properties	$N_A d\overline{H}_A + N_B d\overline{H}_B = 0$	$N_A d\overline{S}_A + N_B d\overline{S}_B = 0$	$N_A d\overline{G}_A + N_B d\overline{G}_B = 0$

This equation is particularly important and called *Gibbs-Duhem equation*.

Solutions

Generalisation for multi-component systems:

$$\overline{Y}_A = \left(\frac{\partial Y}{\partial n_i}\right)_{P,T,n_j} \qquad \sum N_i d\overline{Y}_i = 0 \qquad Y = \sum \overline{Y}_i N_i$$

where $Y = V, U, H, S, A$ or G.

These equations are of value for solution thermodynamics. For example, if a partial molar quantity of one component in a binary solution has been determined, then the partial molar quantity of the other component is fixed:

$$N_A d\overline{Y}_A + N_B d\overline{Y}_B = 0$$

All the general thermodynamic relations can be applied with minor symbolic modifications to the partial molar quantities :

Examples: $\quad G = H - TS \quad \longrightarrow \quad \overline{G}_i = \overline{H}_i + T\overline{S}_i$

$\qquad\qquad\quad dG = VdP - SdT \quad \longrightarrow \quad d\overline{G}_i = \overline{V}_i dP - \overline{S}_i dT$

A graphical method of determining the partial molar quantities from the data on the integral molar quantities is frequently employed.

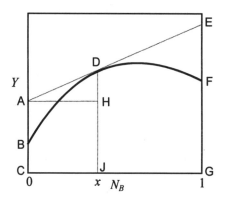

In an A-B two component system, the molar quantity of the solution or the integral molar quantity, Y, at constant P and T is plotted against the mole fraction of B, N_B, in the figure.

$$Y = n_A \overline{Y}_A + n_B \overline{Y}_B$$

Dividing by the total number of moles

$$Y = N_A \overline{Y}_A + N_B \overline{Y}_B$$

Differentiation

$$dY = \overline{Y}_A dN_A + \overline{Y}_B dN_B + N_A d\overline{Y}_A + N_B d\overline{Y}_B$$

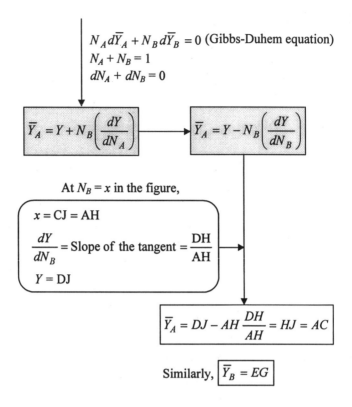

This method is known as *the method of intercepts* and is of proven value in determining partial molar quantities.

Suppose that n_A moles of pure A and n_B moles of pure B are mixed to form a binary solution at constant T and P. The total free energy change, G^M, associated with the mixing may be obtained in the following manner :

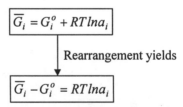

$\bar{G}_i - G_i^o$ is the change in molar free energy of i due to the change in state from the standard state to the state of solution of a particular composition. This is called the *partial molar free energy of mixing* or *relative partial molar free energy* of i and is designated

$$\bar{G}_i^M = \bar{G}_i - G_i^o \longrightarrow \bar{G}_i^M = RT \ln a_i$$

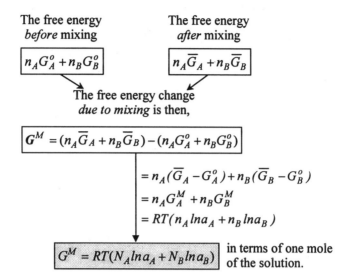

This is called the *relative integral molar free energy or the molar free energy of mixing*. The method of tangential intercepts which we have applied for determination of partial molar quantities from the integral molar quantities can also apply to the relative quantities :

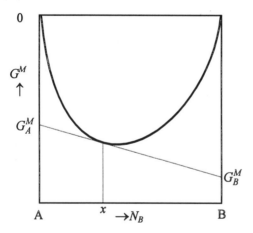

For a solution of specified composition, the relative partial molar entropy can be determined from the data on activities at different temperatures.
Recall that

$$\left(\frac{\partial G}{\partial T}\right)_P = -S \longrightarrow \left(\frac{\partial G_i^M}{\partial T}\right)_P = -S_i^M \longrightarrow S_i^M = -R\ln a_i - RT\left(\frac{\partial (\ln a_i)}{\partial T}\right)_P$$

$$G_i^M = RT \ln a_i$$

Note that data on activities at different temperatures enable us to evaluate the differential in the equation.

Exercises

1. A container having three compartments contains 1 mole of gas A, 2 moles of gas B and 3 moles of gas C, respectively, at the same temperature and pressure (298K and 1 atm). The partitions are lifted and the gases are allowed to mix.
   Calculate the change in the Gibbs free energy G^M. Assume the gases behave ideally.

2. The volume of a dilute solution of KCl of molality m (m moles of KCl in 1 kg of water) is given by the equation

$$V = 1{,}003 + 27.15m + 1.744m^2 \quad (cm^3)$$

   1) Calculate the partial molar volume of KCl ($\overline{V}_{KCl}$) at $m = 0.5$.
   2) Calculate the partial molar volume of KCl at the infinitely dilute solution.

3. The activity coefficient of zinc in liquid zinc-copper alloys in the temperature range of 1,070 to 1,300K can be expresses as follows :

$$RT \ln \gamma_{Zn} = -31{,}600 N_{Cu}^2 \qquad \text{where } R = 8.314 \text{ J mol}^{-1}\text{K}^{-1}$$

   Calculate the activity of Cu in Cu-Zn binary solution of $N_{Cu} = 0.7$ at 1,300K.

2.3. Behaviour of Solutions

2.3.1. Ideal Solutions

We consider the vaporisation of substance A:

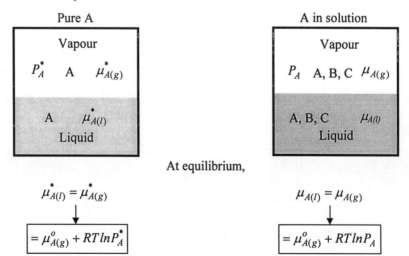

Combination yields,

$$\mu_{A(l)} = \mu^*_{A(l)} + RT\ln\left(\frac{P_A}{P^*_A}\right)$$

In a series of experiments on liquid mixtures, the French scientist F. Raoult found that,

the ratio of the partial vapour pressure of each component of a solution to its vapour pressure as a pure liquid is approximately equal to the mole fraction of the component in the solution.

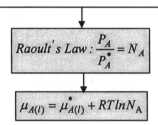

$$\text{Raoult's Law}: \frac{P_A}{P^*_A} = N_A$$

$$\mu_{A(l)} = \mu^*_{A(l)} + RT\ln N_A$$

Solutions that obey these equations throughout the composition range are called *ideal solutions*. Recall that $\mu^*_{A(l)}$ is the chemical potential of pure A at the temperature of interest, but at an arbitrary pressure, and hence is different from $\mu^o_{A(l)}$ which is the chemical potential of pure A at the pressure of 1atm. However, the effect of pressure on the chemical potential is so small that

$$\mu^*_{A(l)} = \mu^o_{A(l)}$$

Recall the definition of activity :

$$a_A = \frac{P_A}{P^*_A}$$

From the Raoult's law, thus,

$$a_A = N_A \quad \text{when A behaves ideally.}$$

Example 1

Prove the following statement:
When components are mixed to form an ideal solution at constant temperature, no heat is absorbed or released.

$$\left[\frac{\partial\left(\frac{G^M_A}{T}\right)}{\partial\left(\frac{1}{T}\right)}\right]_P = H^M_A \qquad G^M_A = RT\ln a_A = RT\ln N_A \quad \longrightarrow \quad H^M_A = 0$$

> **Example 2**
>
> Prove the following statement :
> If species A behaves ideally in the whole composition range of A-B binary solutions, the species B also behaves ideally.

Recall the Gibbs-Duhem equation :

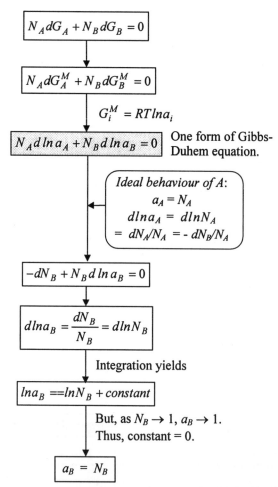

Exercises

1. Liquids A and B are completely miscible and forms an ideal solution. The vapour pressures of two liquids are 2×10^{-3} atm and 5×10^{-3} atm at temperature T, respectively. Calculate the mole fraction of A in liquid phase when the vapour pressure is 4×10^{-3} atm.

2. An ideal solution is made of 79mol% of A, 20mol% of B and 1mol% of C at 298K and 1atm.

   1) Calculate the relative partial molar free energy of A, $\bar{G}_A^M$.
   2) Calculate the relative integral molar free energy of the solution, G^M.
   3) Calculate the relative partial molar entropy of A, $\bar{S}_A^M$.

3. A liquid gold-copper alloy contains 45mol% of copper and behaves ideally at 1,320K. Calculate the amount of heat absorbed in the system when 1g of solid copper is dissolved isothermally at this temperature in a large bath of the alloy of this composition. The following data given :

   $C_{P,Cu(s)} = 22.64 + 6.28 \times 10^{-3} T$, J mol^{-1}K^{-1}
   $C_{P,Cu(l)} = 31.38$ J mol^{-1}K^{-1}
   $\Delta H^o_{f,Cu} = 12,980$ J mol^{-1} (Heat of fusion of Cu)
   $T_{m,Cu} = 1,083°C$ (melting point of Cu)
   $M_{cu} = 63.5$ (atomic weight of Cu)

   Calculate the change in entropy of the system in the above process.

4. A solution is composed of benzene (B) and toluene (T). The Raoult's law holds for both benzene and toluene. The equilibrium vapour pressures of benzene and toluene are 102.4kPa and 39.0kPa, respectively, at 81°C. Calculate the mole fraction of benzene in the vapour which is in equilibrium with the $N_B = 0.5$ solution.

2.3.2. Non-ideal Solutions and Excess Properties

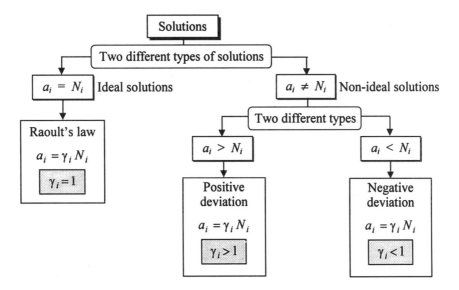

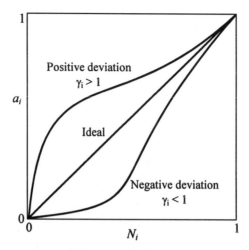

Knowledge of the variation of the value of γ with temperature and composition is of prime importance in solution thermodynamics. Now a question arises as to how to express the extent of deviation of this value from ideal behaviour. As we have discussed in the preceding section, in ideal solutions attractive forces between the unlike molecules in solution are the same as those between the like molecules in the pure phase. Therefore the escaping tendency of the component i in an ideal solution is the same as that in its pure state. From the Raoult's law,

$$P_i = N_i P_i^* \quad \text{for ideal solutions}$$

The positive deviation is characterised by vapour pressures higher than those calculated for ideal solution. If the attraction between the unlike molecules (i-j) is weaker than the mutual attraction of like molecules (i-i or j-j), then the escaping tendencies of the molecules are higher than the escaping tendencies in the individual pure states.

$$\boxed{\left(\frac{P_i}{P_i^*}\right)_{\gamma_i N_i} > \left(\frac{P_i}{P_i^*}\right)_{ideal, N_i}} \longrightarrow \boxed{\gamma_i > 1}$$

The negative deviation, on the other hand, is characterised by vapour pressures lower than those calculated for ideal solution. If the attraction between the unlike molecules (i-j) is stronger than the mutual attraction of the like molecules (i-i or j-j), then the escaping tendencies of the molecules in the solutions are lower than escaping tendencies in the individual liquids. Thus

$$\boxed{\left(\frac{P_i}{P_i^*}\right) < \left(\frac{P_i}{P_i^*}\right)_{ideal}} \longrightarrow \boxed{\gamma_i < 1}$$

We have seen that the properties of non-ideal or real solutions differ from those of ideal solutions. In order to consider the deviation from ideality, we may divide thermodynamic mixing properties into two parts :

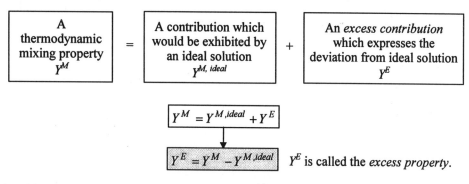

Y^E is called the *excess property*.

Consider the relative partial molar free energy, G_i^M,

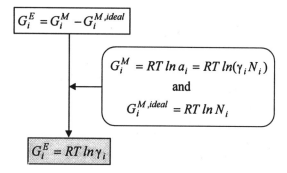

This is called the *excess partial molar free energy* of i. The *excess integral molar free energy* or the *excess molar free energy of solution* is given by

$$G_i^E = \sum(N_i G_i^E) = RT\sum(N_i \ln N_i)$$

In a similar way we may obtain H_i^E and S_i^E:

$$H_i^E = H_i^M \qquad S_i^E = -R\ln\gamma_i - RT\left(\frac{\partial \ln a_i}{\partial T}\right)_P$$

The general thermodynamic relationships, which apply to partial molar quantities, are also valid for the excess quantities.

In general, as the temperature increases, the extent of deviation from ideal behaviour of a non-ideal solution decreases.

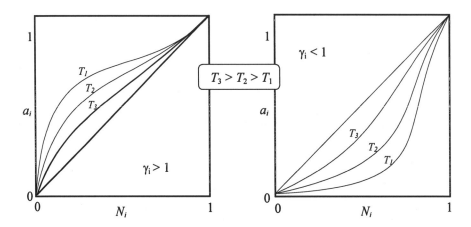

Thus in a solution, if $\gamma_i > 1$, then γ_i decreases as T increases. If $\gamma_i < 1$, then γ_i increases as T increases.

The activity coefficient of a particular component in solution is a measure of the interaction that occurs between atoms or molecules in solution. If $\gamma_i > 1$, then $G_i^E > 0$ (recall $G_i^E = RT \ln\gamma_i$).

Example 1

Prove that, if a component in a solution exhibits positive deviation from ideality, i.e., $\gamma_i > 1$, then the solution process is endothermic.

$$\left[\frac{\partial(G_i^E/T)}{\partial T}\right]_P = -\frac{H_i^E}{T^2} \quad \longrightarrow \quad \left[\frac{\partial(RT \ln\gamma_i)}{\partial T}\right]_P = -\frac{H_i^E}{T^2}$$

$$G_i^E = RT \ln\gamma_i$$

Recall that, if $\gamma_i > 1$, increase in T results in decrease in γ_i. Thus the left-hand side of the above equation becomes negative and hence H_i^E is positive - endothermic. In a similar way one may find that, if $\gamma_i < 1$, then the solution process is exothermic. If the solution process in an *i-j* binary system is endothermic, the *i-i* and *j-j* attractions are greater than the *i-j* attraction. *i* atoms attempt to have only *i* atoms as nearest neighbours - tendency toward *phase separation* or *clustering*.

If the solution process is exothermic, the *i-i* and *j-j* attractions are smaller than the *i-j* attraction. *i* atoms attempt to have only *j* atoms as nearest neighbours, and *j* atoms attempt to have only *i* atoms as nearest neighbours - tendency toward *compound formation*.

Exercises

1. The excess integral molar free energy of the Ga-P binary solution containing up to 50mol% P is

$$G^E = (-7.53T - 2,500)N_P N_{Ga}, \text{ Jmol}^{-1}$$

   Calculate the amount of heat associated with the formation of one mole of solution containing 20mol% P.

2. The integral molar enthalpy and entropy of the Cd-Zn liquid alloy at 432°C are described by the following empirical equations :

$$H^M = 6,700 N_{Cd} N_{Zn} - 1,500 N_{Zn} \ln N_{Zn}, \quad \text{Jmol}^{-1}$$

$$S^M = -8.4(N_{Cd} \ln N_{Cd} + N_{Zn} \ln N_{Zn}) \quad \text{Jmol}^{-1}\text{K}^{-1}$$

   Calculate the excess partial molar free energy of cadmium, G_{Cd}^E, at $N_{cd} = 0.5$.

2.3.3. Dilute Solutions

When the following relationship holds:

Vapour pressure exerted by component i in a solution (P_i)	=	Mole fraction of component i in the solution (N_i)	x	Vapour pressure of i in equilibrium with pure i (P_i^*)

$$\boxed{P_i = N_i P_i^*}$$

then we say that the solute i in the solution obeys *Raoult's law* and *behaves ideally*. The ideal behaviour is graphically represented in the following figure :

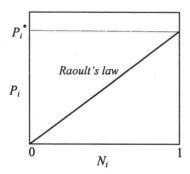

In real solution, on the other hand, the vapour pressure of a solute deviates from Raoult's law.

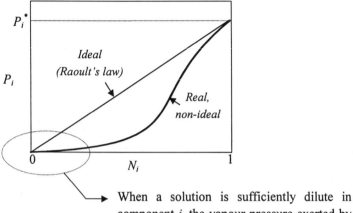

When a solution is sufficiently dilute in component i, the vapour pressure exerted by component i is linearly proportional to the concentration of component i.

This relationship is known as *Henry's Law*.

$$P_i = k_i N_i$$

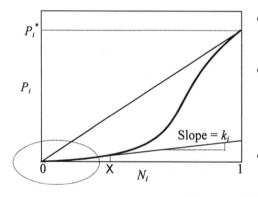

- Note that k_i is the slope of the tangent drawn from the zero concentration.
- Note also that, if $k_i = P_i^*$, then Henry's law becomes identical to Raoult's law. Thus, Henry's law may be considered more general than Raoult's law.
- From the figure we can see that the component i obeys Henry's law up to the mole fraction of x.

Summary of Raoult's law and Henry's law:

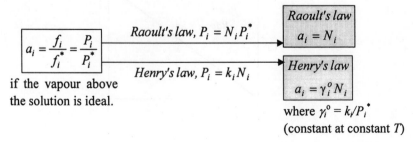

$$a_i = \frac{f_i}{f_i^*} = \frac{P_i}{P_i^*}$$

if the vapour above the solution is ideal.

Raoult's law, $P_i = N_i P_i^*$

Raoult's law
$a_i = N_i$

Henry's law, $P_i = k_i N_i$

Henry's law
$a_i = \gamma_i^o N_i$

where $\gamma_i^o = k_i/P_i^*$
(constant at constant T)

Solutions

In a binary solution, if the solute follows Henry's law, the solvent follows Raoult's law. (One may prove this using Gibbs-Duhem equation.)

Exercises

1. The following table shows the vapor pressures of A exerted by A-B alloys at 1,273K :

N_A	0.1	0.2	0.3	0.4	0.5	0.6	0.7	0.8	0.9	1.0
$P_a \times 10^6$, atm	0.25	0.5	0.75	01.2	2.0	2.85	3.75	4.8	5.4	6.0

Up to what mole fraction does the solute A obey Henry's law?

2. The activity of carbon, a_C, in liquid Fe-C alloys is given by the following equation:

$$\log a_C = \log\left(\frac{N_C}{1-2N_C}\right) + \frac{1,180}{T} - 0.87 + \left(0.72 + \frac{3,400}{T}\right)\left(\frac{N_C}{1-N_C}\right)$$

where the standard state of carbon is pure graphite.

1) Find Henrian activity coefficient, γ_C^o of liquid Fe-C solutions at 1,600°C.
2) Calculate H_C^M in the composition range over which carbon obeys Henry's law.

2.3.4. Gibbs-Duhem Equation

In Section 2.2.2. we have developed the following general form of the *Gibbs-Duhem equation* :

$$\boxed{\sum N_i d\overline{Y}_i = 0}$$

Applying to A-B binary solution for partial molar free energies,

$$\boxed{N_A d\overline{G}_A + N_B d\overline{G}_B = 0}$$

$$G_i^M = \overline{G}_i - G_i^o$$
$$dG_i^M = d\overline{G}_i \text{ at constant } T$$

$$\boxed{N_A dG_A^M + N_B dG_B^M = 0}$$

$$G_i^M = RT \ln a_i$$

$$N_A d\ln a_A + N_B d\ln a_B = 0$$

$$a_i = \gamma_i N_i$$
$$N_A + N_B = 1 \rightarrow dN_A + dN_B = 0$$

$$N_A d\ln \gamma_A + N_B d\ln \gamma_B = 0$$

The last two equations are particularly useful in the determination of activities. If the activity of one component is known over a range of compositions, then it is possible to determine the activity of the other component by employing either one of the equations. Suppose that the activities of component B are known over a range of compositions.

$$N_A d\ln a_A + N_B d\ln a_B = 0$$

Rearrangement and integration

$$\int_{a_A=1}^{a_A=a_A} d\ln a_A = \int_{a_A=1}^{a_A=a_A} -\frac{N_B}{N_A} d\ln a_B$$

$$a_A = a_A \rightarrow N_A = N_A$$
$$a_A = 1 \rightarrow N_A = 1$$

$$\int_{a_A=1}^{N_A=N_A} d\ln a_A = \int_{N_A=1}^{N_A=N_A} -\frac{N_B}{N_A} d\ln a_B$$

$$\ln a_A \Big|_{N_A=N_A} = \int_{N_A=1}^{N_A=N_A} -\frac{N_B}{N_A} d\ln a_B$$

This equation can be solved by graphical integration as shown below:

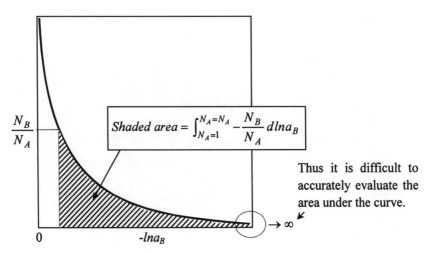

Thus it is difficult to accurately evaluate the area under the curve.

Now, let's consider the other form of the Gibbs-Duhem equation which involves activity coefficients in place of activities :

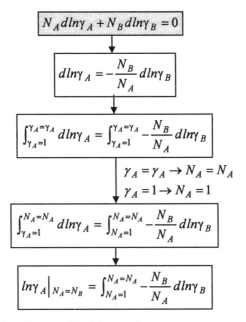

This equation can be solved by graphical integration.

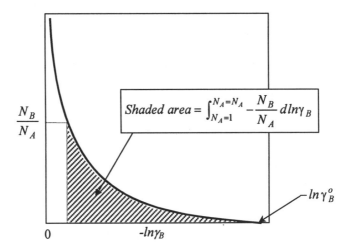

With this method, thus, it is possible to accurately evaluate the area under the curve.

Darken and Gurry have introduced a new function called the *alpha function*, α :

$$\boxed{\alpha_i = \frac{\ln \gamma_i}{(1-N_i)^2}} \xrightarrow[\text{Gibbs-Duhem equation}]{\text{Combination with}} \boxed{\ln \gamma_A = -\alpha_B N_A N_B - \int_{N_A=1}^{N_A=N_A} \alpha_B dN_A}$$

where i = A or B.

This equation enables us to calculate γ_A at $N_A = N_A$ by graphically integrating α_B values over the range from $N_A = 1$ to $N_A = N_A$.

It is obvious from the equation that the activity coefficient of A at infinite dilution of A, i.e., the Henrian activity coefficient, γ_A^o is given by

$$ln\gamma_A^o = \int_{N_A=0}^{N_A=1} \alpha_B dN_A$$

Exercises

1. The activity of zinc in liquid cadmium-zinc alloys at 708K is related to the alloy composition by the following equation :

$$ln\gamma_{Zn} = 0.87 N_{Cd}^2 - 0.3 N_{Cd}^3$$

Calculate the activity of cadmium at $N_{cd} = 0.1$.

2. The relative integral molar enthalpy or the molar heat of mixing of liquid Sn-Bi solutions at 330°C is represented by

$$H^M = 400 N_{Sn} N_{Bi} \, \text{Jmol}^{-1}$$

Determine H_{Sn}^M at $N_{sn} = 0.3$.

2.3.5. Solution Models

The activity of component i in a solution is related to the concentration of i in the solution and determination of the variation of γ_i with T and composition is of critical importance in chemical thermodynamics.

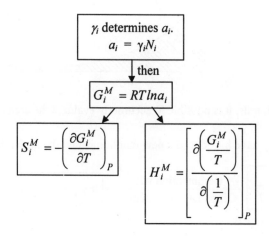

Several formalisms have been developed to relate the activity coefficient of a component in the solution to the temperature and composition of the solution. We now discuss the following two models :
- Regular solution model
- Margules formalism

Regular Solutions

Some solutions exhibit that mixing is random (ideal solution), but net heat absorbed or released is not zero ($H^M \neq 0$: non-ideal solution). Solutions, which exhibit the behaviour above-mentioned, are called *regular solutions*.

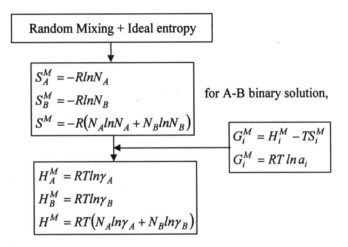

for A-B binary solution,

Hildebrand[*] showed that in the regular binary A-B solution,

$$\boxed{RT \ln \gamma_A = \Omega N_B^2} \qquad \boxed{RT \ln \gamma_B = \Omega N_A^2}$$

Ω is called the *interaction parameter* and independent of composition and temperature. Let's examine the properties of a regular solution using the concept of excess properties.

$$G_A^E = G_A^M - G_A^{M,ideal} = RT \ln \gamma_A$$
$$G_B^E = G_B^M - G_B^{M,ideal} = RT \ln \gamma_B$$

Comparing the excess free energies with the excess enthalpies, we obtain

$$\boxed{G_i^E = H_i^E = RT \ln \gamma_i} \text{ for regular solutions}$$

---

[*] J.H. Hildebrand, J. Amer. Chem. Soc., 51, 1929, p66-80

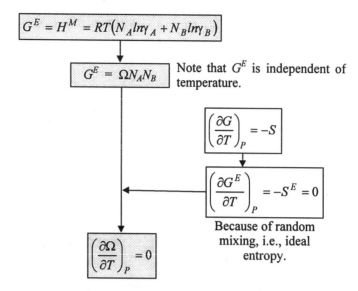

The interaction parameter, Ω, is thus proved to be independent of temperature.

For regular solutions, activity at one temperature can be calculated, if activity at other temperature is known:

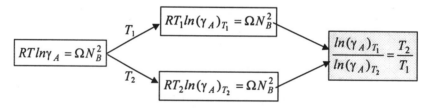

Margules Formalism

Margules suggested a power series formula for expressing the activity-composition variation of a binary solution :

$$ln\gamma_A = \alpha_1 N_B + \frac{1}{2}\alpha_2 N_B^2 + \frac{1}{3}\alpha_3 N_B^3 + \cdots$$

$$ln\gamma_B = \beta_1 N_A + \frac{1}{2}\beta_2 N_A^2 + \frac{1}{3}\beta_3 N_A^3 + \cdots$$

Applying the Gibbs-Duhem equation with ignoring coefficients α_i's and β_i's higher than $i = 3$, we can obtain

$$\alpha_1 = \beta_1 = 0, \quad \beta_2 = \alpha_2 + \alpha_3, \quad \beta_3 = -\alpha_3$$

Then

$$ln\gamma_A = \frac{1}{2}\alpha_2 N_B^2 + \frac{1}{3}\alpha_3 N_B^3 \qquad ln\gamma_B = \frac{1}{2}(\alpha_2 + \alpha_3)N_A^2 - \frac{1}{3}\alpha_3 N_A^3$$

or

$$\ln\gamma_A = N_B^2[A-2N_A(A-B)] \qquad \ln\gamma_B = N_A^2[A-2N_B(A-B)]$$

where, $A = 1/2\, \alpha_2 + 1/3\, \alpha_3$ and $B = 1/2\, \alpha_2 + 1/6\, \alpha_3$

These equations are known as *three suffix Margules equations*.

Exercises

1. Zinc and cadmium liquid alloys conform to regular solution behaviour. The following table shows the relative integral molar enthalpies (or the molar heats of mixing) at various zinc concentrations at 723K.

N_{Zn}	0.06	0.09	0.15	0.37	0.61	0.76	0.86	0.95
H^M(J/mol)	493	714	1,126	1,985	2,030	1,585	1,039	423

   1) Calculate the average interaction parameter Ω.
   2) Calculate the activity of Zn in the solution containing $N_{zn} = 0.3$.

2. In the A-B binary solution the activity coefficients are given by the three-suffix Margules equations :

   $$\ln\gamma_A = \frac{1}{2}\alpha_2 N_B^2 + \frac{1}{3}\alpha_3 N_B^3$$

   $$\ln\gamma_B = \frac{1}{2}(\alpha_2 + \alpha_3)N_A^2 - \frac{1}{3}\alpha_3 N_A^3$$

   It was found that the Henrian activity coefficients of A and B at the temperature T, γ_A^o and γ_B^o were 0.75 and 0.54, respectively. Determine α_2.

CHAPTER 3

EQUILIBRIA

3.1. Reaction Equilibria

3.1.1 Equilibrium Constant

Consider a general chemical reaction occurring at constant temperature and pressure :

$$aA + bB + \cdots = mM + nN + \cdots$$

where $a, b, ..., m, n, ...$ are stoichiometric coefficients indicating the number of moles of respective species A, B, ..., M, N, ...

Let ΔG_r denote the free energy change associated with the above reaction.

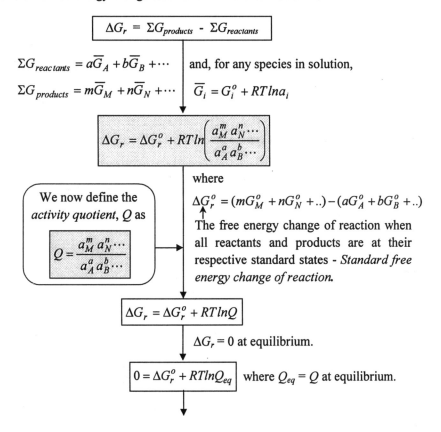

$$Q_{eq} = \exp\left(-\frac{\Delta G_r^o}{RT}\right)$$

Let $K = Q_{eq}$

$$K = \exp\left(-\frac{\Delta G_r^o}{RT}\right)$$ K is called *equilibrium constant*.

$$K = \left(\frac{a_M^m a_N^n \cdots}{a_A^a a_B^b \cdots}\right)_{eq}$$

Note that the values of a_i's in this expression for K are the values at equilibrium, and hence ΔG_r becomes zero when these values are plugged into the equation for ΔG_r.

- Note that since a_i's are dimensionless, both K and Q are also dimensionless.
- Note that K can be determined from the knowledge of either ΔG_r^o or a_i's at equilibrium, whereas Q has nothing to do with ΔG_r^o.
- We may obtain the following equation by proper combination of the equations for ΔG_r, Q and K:

$$\Delta G_r = RT \ln\left(\frac{Q}{K}\right)$$

Note that
if $Q/K < 1$ → the reaction is spontaneous from left to right as written,
if $Q/K = 1$ → the reaction is at equilibrium, and
if $Q/K > 1$ → the reaction is spontaneous from right to left as written.

Example 1

The figure given below depicts some reaction paths that the following general reaction may take:

$$aA + bB = mM + nN \quad \Delta G_r$$

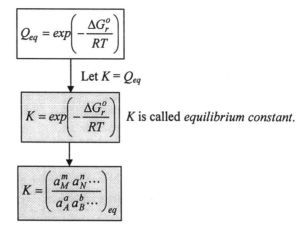

Prove the following relationships:

$$\Delta G_1 = -RT \ln\left(a_A^a a_B^b\right) \quad \Delta G_2 = -RT \ln\left(a_M^m a_N^n\right) \quad \Delta G_r = \Delta G_r^o + RT \ln\left(\frac{a_M^m a_N^n}{a_A^a a_B^b}\right)$$

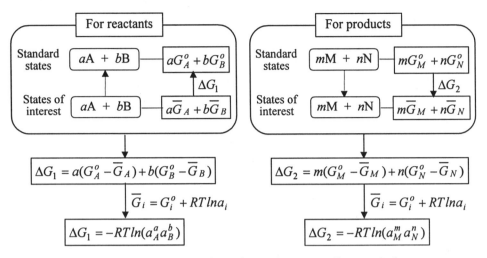

The free energy change of reaction at the states of interest is then

$$\Delta G_r = \Delta G_1 + \Delta G_r^o + \Delta G_2$$

$$\Delta G_r = \Delta G_r^o + RT \ln\left(\frac{a_M^m a_N^n}{a_A^a a_B^b}\right)$$

This example clearly shows the distinction between ΔG_r and ΔG_r^o.

Example 2

Is the following statement true or false?
Since the equilibrium constant K does not have units, one does not have to be concerned with units in the formulation of K.

Consider the following simple reaction involving a gas species :

$$2A(c) + B(g) = M(c)$$

where c = condensed phase, and g = gaseous phase.

$$K = \frac{a_M}{a_A^2 a_B}$$

$$a_B = \frac{f_B}{f_B^o} \cong \frac{P_B}{P_B^o}$$

$$K = \frac{a_M}{a_A^2 P_B}$$

If we choose the 1atm ideal gas standard state for gas B, $P_B^o = 1(atm)$.

Thus, it is customary to express activities of gaseous species in terms of the pressure symbol P. It has to be remembered that P_i in the formulation of K is in fact P_i/P_i^o and hence has the same numerical value as the pressure in *atm* unit of the species i only if the 1atm ideal gas standard state is chosen for species i.

Exercises

1. Consider the following reaction at 873K :

$$2SO_2(g) + O_2(g) = 2SO_3(g)$$

1) Calculate the standard free energy change of reaction (ΔG_r^o) at 873K. The following data are given :

$$\Delta G_{f,SO_2}^o = -361,670 + 72.68T, J$$
$$\Delta G_{f,SO_3}^o = -457,900 + 163.34T, J$$

2) Calculate the equilibrium constant K of the reaction at 873K.

3) A gas mixture includes SO_2, SO_3 and O_2. Partial pressures of these species are given in the following :

$$P_{SO_2} = 0.01 atm, \quad P_{SO_3} = 0.1 atm, \quad P_{O_2} = 0.21 atm.$$

Calculate the activity quotient Q.

4) Calculate the free energy change of the reaction, ΔG_r.

5) Is this reaction spontaneous from left to right as written under the conditions given above?

6) If the partial pressures are changed as indicated below, calculate the free energy change of the reaction:

$$P_{SO_2} = 0.01 atm, \quad P_{SO_3} = 0.1 atm, \quad P_{O_2} = 0.21 atm.$$

7) Is this reaction spontaneous from left to right as written?

8) Calculate the partial pressure of SO_3 which would be in equilibrium at 873K with

$$P_{SO_2} = 0.01 atm, \quad P_{O_2} = 0.21 atm.$$

2. Excess amount of pure carbon is reacted with $H_2O(g)$ at 1,000°C.

$$C(s) + H_2O(g) = CO(g) + H_2(g)$$

The gas phase does not contain any other species and is maintained at 1atm of the total pressure. Calculate the equilibrium partial pressure of $H_2O(g)$. The following data are given :

$$C(s) + \tfrac{1}{2}O_2(g) = CO(g) \quad \Delta G_1^o = -112,880 - 86.51T, Jmol^{-1}$$
$$H_2(g) + \tfrac{1}{2}O_2(g) = H_2O(g) \quad \Delta G_2^o = -247,390 + 55.85T, Jmol^{-1}$$

3.1.2. Criteria of Reaction Equilibrium

Prior to deriving conditions for equilibrium of a system, mathematics on the extremum principle is revisited in the following:

Extremum Principle

Consider a function g of a variable x,

$$g = g(x)$$

An extreme value of g can be found by differentiating the equation, and the extreme values occur at the points on the curve at which the derivative is zero;

$$\frac{dg}{dx} = 0$$

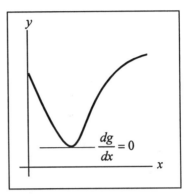

Next, consider a function g of two independent variables x and y:

$$g = g(x,y) \quad \xrightarrow{\text{Differentiation yields}} \quad dg = \left(\frac{\partial g}{\partial x}\right)_y dx + \left(\frac{\partial g}{\partial y}\right)_x dy$$

Referring to the figure below, at the minimum,

$$\left(\frac{\partial g}{\partial x}\right)_y = 0 \quad \text{and} \quad \left(\frac{\partial g}{\partial y}\right)_x = 0$$

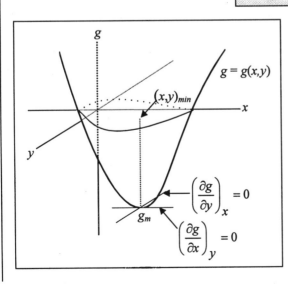

These two equations determine the extremum of the function g. The values of x and y at the extremum can be found by solving the two equations simultaneously.

In general, for a function g,

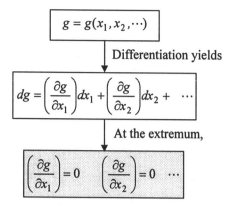

For a function $g = g(x,y)$, suppose y is not independent of x, but related to x as

$$y = y(x)$$

In this case, as shown in the following figure, the extremum with this constraint, $g_{m(c)}$, may differ from the unconstrained absolute extremum, g_m:

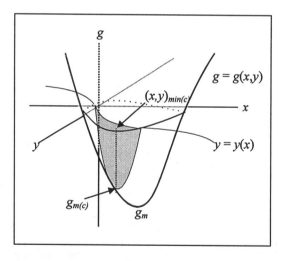

As an example, suppose g is a function of x and y as given below :

$$g = x^2 + y^2 - 2x - 6y + 14$$

Differentiation yields

$$dg = 2xdx + 2ydy - 2dx - 6dy = (2x - 2)dx + (2y - 6)dy$$

For the absolute extremum (minimum in this case),

$$2x - 2 = 0 \quad \text{and} \quad 2y - 6 = 0, \quad \text{Thus} \quad x = 1, \text{ and } y = 3.$$

If we have a constraint of

$$y = 2x$$

Differentiation yields

$$dy = 2dx$$

Then the equation above for dg becomes

$$dg = 2xdx + 2(2x)(2dx) - 2dx - 6(2dx) = (10x - 14)dx$$

The *constrained extremum* thus occurs at $10x - 14 = 0$ or $x = 1.4$ and $y = 2.8$.
The above example clearly shows the difference between the absolute extremum and the constrained extremum.

A system has the minimum value of the Gibbs free energy at equilibrium. Suppose we have a system which consists of three phases, namely gas, liquid and solid solutions, with a number of different species (1, 2, 3, ...):

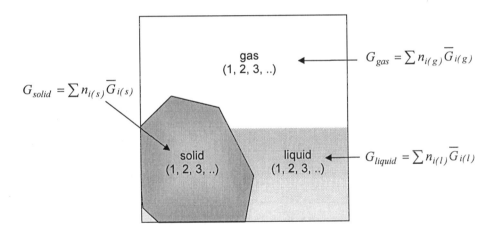

The total Gibbs free energy of the system (G_{system})

$$G_{system} = G_{gas} + G_{liquid} + G_{solid}$$

$$G_{system} = \sum n_{i(g)} \overline{G}_{i(g)} + \sum n_{i(l)} \overline{G}_{i(l)} + \sum n_{i(s)} \overline{G}_{i(s)}$$

$$\overline{G}_i = G_i^o + RT \ln a_i$$

$$G_{system} = \left(\sum n_{i(g)} G_{i(g)}^o + \sum n_{i(l)} G_{i(l)}^o + \sum n_{i(s)} G_{i(s)}^o \right) + RT \left(\sum n_{i(g)} \ln a_{i(g)} + \sum n_{i(l)} \ln a_{i(l)} + \sum n_{i(s)} \ln a_{i(s)} \right)$$

At equilibrium, the total Gibbs free energy of the system, G_{system}, is at minimum: i.e.,

$$\left(\frac{\partial G_{system}}{\partial n_{i(k)}}\right)_{n_j} = 0 \quad \text{where } i : \text{component 1, 2, ...} \\ k : \text{phase } g, l, s$$

but subject to two constraints :
- The atom balance relations must be satisfied.
- n_i and a_i have non-negative values.

Solving the above set of simultaneous equations subject to the two constraints, one can find, at least in principle, distribution of each component between phases at equilibrium ($n_{i(k)}$). The following diagram shows a two dimensional graphical interpretation of the equilibrium conditions discussed above :

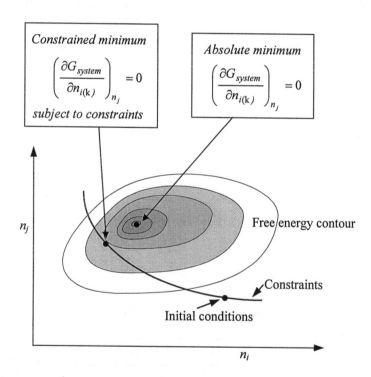

Example 1

In a reactor $CO(g)$ and $O_2(g)$ were introduced and allowed to react to form $CO_2(g)$ at constant temperature T and 1 atm pressure. At equilibrium it was found that all three gaseous species coexisted together. Find the equilibrium conditions of the system by utilising the fact that the Gibbs free energy of the system is minimum at equilibrium.

Equilibria

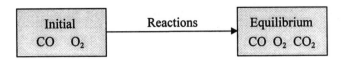

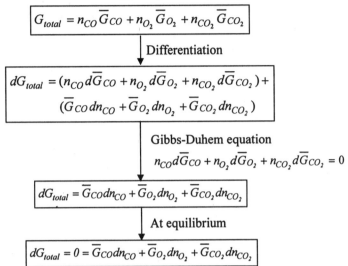

As chemical reactions are allowed to occur in the system, the number of moles of the components are not conserved. Even if the process proceeds in a closed system which does not allow matter to cross the system boundary, the number of moles of each of the components may still vary: i.e., a component may be either consumed or produced by reactions. However, the number of gram atoms of each of the elements in the system must be conserved as atoms cannot be created or destroyed. This fact imposes constraints to the above equation.

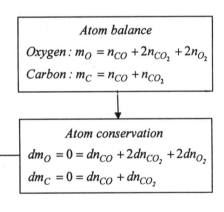

This equation is significant. Notice that the left-hand side of the equation is in fact the change in the free energy for the reaction

$$CO + \tfrac{1}{2}O_2 = CO_2$$

This free energy change is zero as shown in the above equation, and hence there is no driving force for the reaction to proceed in either direction. This means that the above reaction is at equilibrium.

Using the relationship
$$\overline{G}_i = G_i^o + RT \ln P_i$$

$$G_{CO_2}^o - \left(G_{CO}^o + \tfrac{1}{2}G_{O_2}^o\right) + RT \ln \left(\frac{P_{CO_2}}{P_{CO} P_{O_2}^{\frac{1}{2}}}\right) = 0$$

$$\frac{P_{CO_2}}{P_{CO} P_{O_2}^{\frac{1}{2}}} = \exp\left(-\frac{\Delta G^o}{RT}\right)$$

where $\Delta G^o = G_{CO_2}^o - (G_{CO}^o + \tfrac{1}{2}G_{O_2}^o)$

The above equation determines the relationship between the partial pressures of CO, CO$_2$ and O$_2$ at equilibrium. The analysis given above is the basis of the concept of *equilibrium constant* discussed in the prior section.

Example 2

Pure FeO(s) is reduced to Fe(s) by CO(g) in a reactor at constant temperature T and 1atm pressure. At equilibrium, Fe(s) and FeO(s) coexist with CO(g) and CO$_2$(g). Relate the equilibrium ratio of CO$_2$/CO to the standard free energies of formation of species existing in the system and to the temperature T.

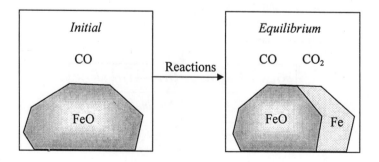

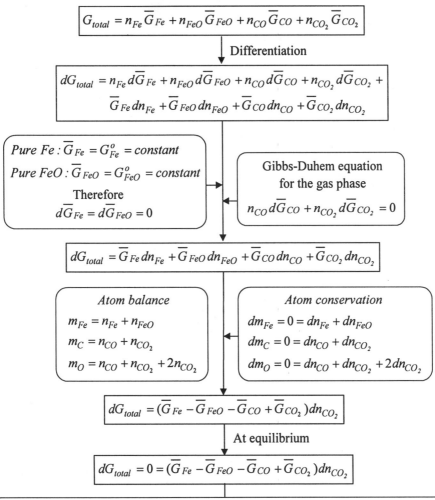

This equation shows that the free energy change associated with the following reaction is zero :

$$FeO + CO = Fe + CO_2$$

In other words the reaction is at equilibrium, which is consistent with the imposed condition of $dG_{total} = 0$.

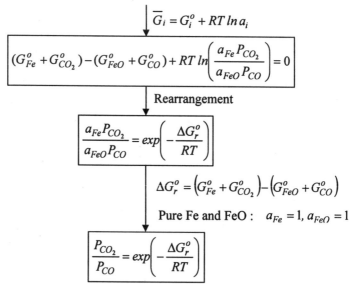

Example 3

In a steel refining process, molten steel eventually comes into equilibrium with slag and gas phases coexisting in the furnace. The furnace can be considered as a closed system. The species identified in each phase are given in the following:

> Metal phase : Fe, Si, Mn, C, O, N, S, Al
> Slag phase : CaO, SiO$_2$, MnO, FeO, CaS, Al$_2$O$_3$
> Gas phase : N$_2$, CO

Determine thermodynamic relationships between these species by using the *free energy minimisation method*, i.e., the fact that the free energy is minimum at equilibrium.

Gas
N$_2$, CO

$$G_{gas} = n_{N_2}\overline{G}_{N_2} + n_{CO}\overline{G}_{CO}$$

Slag
CaO, SiO$_2$, MnO, FeO, CaS, Al$_2$O$_3$

$$G_{slag} = n_{CaO}\overline{G}_{CaO} + n_{SiO_2}\overline{G}_{SiO_2} + n_{MnO}\overline{G}_{MnO} + n_{FeO}\overline{G}_{FeO} + n_{CaS}\overline{G}_{CaS} + n_{Al_2O_3}\overline{G}_{Al_2O_3}$$

Metal
Fe, Si, Mn, C, O, N, S, Al

$$G_{metal} = n_{Fe}\overline{G}_{Fe} + n_{Si}\overline{G}_{Si} + n_{Mn}\overline{G}_{Mn} + n_C\overline{G}_C + n_O\overline{G}_O + n_N\overline{G}_N + n_S\overline{G}_S + n_{Al}\overline{G}_{Al}$$

Equilibria

$$G_{total} = G_{gas} + G_{slag} + G_{metal}$$

Differentiation

$$dG_{total} = dG_{gas} + dG_{slag} + dG_{metal}$$

Substituting G_{gas}, G_{slag}, and G_{metal} with the equations given above, and applying Gibbs-Duhem equation,

$$dG_{total} = \overline{G}_{N_2} dn_{N_2} + \overline{G}_{CO} dn_{CO} + \overline{G}_{CaO} dn_{CaO} + \overline{G}_{SiO_2} dn_{SiO_2} + \overline{G}_{MnO} dn_{MnO} +$$
$$\overline{G}_{FeO} dn_{FeO} + \overline{G}_{CaS} dn_{CaS} + \overline{G}_{Al_2O_3} dn_{Al_2O_3} + \overline{G}_{Fe} dn_{Fe} + \overline{G}_{Si} dn_{Si} +$$
$$\overline{G}_{Mn} dn_{Mn} + \overline{G}_C dn_C + \overline{G}_O dn_O + \overline{G}_N dn_N + \overline{G}_S dn_S + \overline{G}_{Al} dn_{Al}$$

Atom balances

$dm_N = 0 = 2dn_{N_2} + dn_N$

$dm_C = 0 = dn_{CO} + dn_C$

$dm_O = 0 = dn_{CO} + dn_{CaO} + 2dn_{SiO_2} + dn_{MnO} + dn_{FeO} + 3dn_{Al_2O_3} + dn_O$

$dm_S = 0 = dn_{CaS} + dn_S$

$dm_{Fe} = 0 = dn_{FeO} + dn_{Fe}$

$dm_{Ca} = 0 = dn_{CaO} + dn_{CaS}$

$dm_{Mn} = 0 = dn_{MnO} + dn_{Mn}$

$dm_{Si} = 0 = dn_{SiO_2} + dn_{Si}$

$dm_{Al} = 0 = 2dn_{Al_2O_3} + dn_{Al}$

$$dG_{total} = (\overline{G}_N - \tfrac{1}{2}\overline{G}_{N_2}) dn_N + (\overline{G}_{CO} - \overline{G}_C - \overline{G}_O) dn_{CO} +$$
$$(\overline{G}_{CaS} + \overline{G}_O - \overline{G}_{CaO} - \overline{G}_S) dn_{CaS} + (\overline{G}_{SiO_2} - \overline{G}_{Si} - 2\overline{G}_O) dn_{SiO_2} +$$
$$(\overline{G}_{MnO} - \overline{G}_{Mn} - \overline{G}_O) dn_{MnO} + (\overline{G}_{FeO} - \overline{G}_{Fe} - \overline{G}_O) dn_{FeO} +$$
$$(\overline{G}_{Al_2O_3} - 2\overline{G}_{Al} - 3\overline{G}_O) dn_{Al_2O_3}$$

At equilibrium, $dG_{total} = 0$ and dn_i's $\neq 0$

$0 = \overline{G}_N - \tfrac{1}{2}\overline{G}_{N_2}$

$0 = \overline{G}_{CO} - (\overline{G}_C + \overline{G}_O)$

$0 = \overline{G}_{CaS} + \overline{G}_O - (\overline{G}_{CaO} + \overline{G}_S)$

$0 = \overline{G}_{MnO} - (\overline{G}_{Mn} + \overline{G}_O)$

$0 = \overline{G}_{SiO_2} - (\overline{G}_{Si} + 2\overline{G}_O)$

$0 = \overline{G}_{FeO} - (\overline{G}_{Fe} + \overline{G}_O)$

$0 = \overline{G}_{Al_2O_3} - (2\overline{G}_{Al} + 3\overline{G}_O)$

Notice that each of these reactions indicates that the free energy change of the corresponding reaction must be zero: i.e., the reaction is at equilibrium.

1/2 N_2 = N
C + O = CO
CaO + S = CaS + O
Mn + O = MnO
Si + 2O = SiO_2
Fe + O = FeO
2Al + 3O = Al_2O_3

The above analysis results in an important conclusion : If a system is at equilibrium, all subsystems in the system must also be at equilibrium.

Combining each of the above equations with the equation $\bar{G}_i = G_i^o + RT \ln a_i$

$$\frac{a_N}{P_{N_2}^{\frac{1}{2}}} = \exp\left(-\frac{\Delta G_N^o}{RT}\right) \qquad \text{where } \Delta G_N^o = G_N^o - \tfrac{1}{2}G_{N_2}^o$$

$$\frac{P_{CO}}{a_C a_O} = \exp\left(-\frac{\Delta G_{CO}^o}{RT}\right) \qquad \text{where } \Delta G_{CO}^o = G_{CO}^o - G_C^o - G_O^o$$

$$\frac{a_{CaS} a_O}{a_{CaO} a_S} = \exp\left(-\frac{\Delta G_{CaS}^o}{RT}\right) \qquad \text{where } \Delta G_{CaS}^o = G_{CaS}^o + G_O^o - G_{CaO}^o - G_S^o$$

$$\frac{a_{MnO}}{a_{Mn} a_O} = \exp\left(-\frac{\Delta G_{MnO}^o}{RT}\right) \qquad \text{where } \Delta G_{MnO}^o = G_{MnO}^o - G_{Mn}^o - G_O^o$$

$$\frac{a_{SiO_2}}{a_{Si} a_O^2} = \exp\left(-\frac{\Delta G_{SiO_2}^o}{RT}\right) \qquad \text{where } \Delta G_{SiO_2}^o = G_{SiO_2}^o - G_{Si}^o - 2G_O^o$$

$$\frac{a_{FeO}}{a_{Fe} a_O} = \exp\left(-\frac{\Delta G_{FeO}^o}{RT}\right) \qquad \text{where } \Delta G_{FeO}^o = G_{FeO}^o - G_{Fe}^o - G_O^o$$

$$\frac{a_{Al_2O_3}}{a_{Al}^2 a_O^3} = \exp\left(-\frac{\Delta G_{Al_2O_3}^o}{RT}\right) \qquad \text{where } \Delta G_{Al_2O_3}^o = G_{Al_2O_3}^o - 2G_{Al}^o - 3G_O^o$$

- If the system is at equilibrium, all of the above equations must be satisfied simultaneously.
- Each of the equations is the condition for equilibrium of the corresponding reaction which occurs in the system.
- Notice that the number of equations, i.e., the number of independent reactions (r) is related with the number of components (c) and the number of elements (e) by the equation

$$r = c - e$$

For instance, for the system under discussion,
c = 16 (N_2, CO, CaO, SiO_2, MnO, FeO, CaS, Al_2O_3, Fe, Si, Mn, C, O, N, S, Al)
e = 9 (N, C, O, Ca, Si, Mn, Fe, S, Al)
r = 16 - 9 = 7
More discussion with rigor is given in the section 3.2.1.

Exercises

1. A system consisting of ZnO(s), C(s), CO(g), Zn(g), CO$_2$(g) and O$_2$ is at equilibrium at 1200K and 1atm. Using the free energy minimisation method, calculate the equilibrium partial pressures for gaseous species in the system. The following data are given :

$$Zn(g) + \tfrac{1}{2}O_2(g) = ZnO(s) \quad \Delta G^o = -460,240 + 198.32T, Jmol^{-1}$$
$$C(s) + \tfrac{1}{2}O_2(g) = CO(g) \quad \Delta G^o = -112,880 - 86.51T, Jmol^{-1}$$
$$C(s) + O_2(g) = CO_2(g) \quad \Delta G^o = -394,760 - 0.836T, Jmol^{-1}$$

2. Excess K$_2$CO$_3$ and C are heated together in an initially evacuated vessel to a temperature of 1,400K and allowed to reach equilibrium. No liquid or solid phases are formed. The gas phase contains K, CO and CO$_2$. Calculate the partial pressures of these species at 1,440K. Use the free energy minimisation method. The following data are given :

Species	K	CO	CO$_2$	K$_2$CO$_3$
ΔG_f^o (Jmol^{-1})	0	-235,100	-396,300	-708,700

3. A reducing gas mixture consisting of 24 mol% CO, 4 mol% CO$_2$, 60 mol% H$_2$ and 12 mol% H$_2$O is passed through a packed bed of wustite (FeO) pellets held at 1,100K. The pressure is maintained constant at 2.8atm. Assuming that the gas phase and the solids are in equilibrium, calculate the composition of the exit gas phase using the free energy minimisation method. The following data are given :

$$C(s) + \tfrac{1}{2}O_2(g) = CO(g) \quad \Delta G^o = -112,880 - 86.51T, Jmol^{-1}$$
$$C(s) + O_2(g) = CO_2(g) \quad \Delta G^o = -394,760 - 0.836T, Jmol^{-1}$$
$$Fe(s) + \tfrac{1}{2}O_2(g) = FeO(s) \quad \Delta G^o = -264,000 + 64.59T, Jmol^{-1}$$
$$H_2(g) + \tfrac{1}{2}O_2(g) = H_2O(g) \quad \Delta G^o = -247,390 + 55.85T, Jmol^{-1}$$

3.1.3. Effect of Temperature on Equilibrium Constant

How will the position of equilibrium change when the temperature is altered?
We can answer this question by using the thermodynamic relations we have developed so far.

Recall that

$$\left[\frac{\partial\left(\frac{G}{T}\right)}{\partial\left(\frac{1}{T}\right)}\right]_P = H$$

Applying to a chemical reaction,

$$\left[\frac{\partial\left(\frac{\Delta G_r^o}{T}\right)}{\partial\left(\frac{1}{T}\right)}\right]_P = \Delta H_r^o$$

At equilibrium

$$K = \exp\left(-\frac{\Delta G_r^o}{RT}\right) \text{ or } \Delta G_r^o = -RT\ln K$$

$$\left(\frac{\partial \ln K}{\partial T}\right)_P = \frac{\Delta H_r^o}{RT^2}$$

This equation is known as *the van't Hoff equation*, and expresses the temperature dependence of the equilibrium constant in terms of the heat of reaction. This equation tells us that

- If the reaction is endothermic, i.e., $\Delta H_r^o > 0$, K increases with increasing T,
- If the reaction is exothermic, i.e., $\Delta H_r^o < 0$, K decreases with increasing T.

If ΔH_r^o is independent* of T, the integration of the van't Hoff equation yields

$$\ln\left(\frac{K_2}{K_1}\right) = \frac{\Delta H_r^o}{R}\left(\frac{1}{T_1} - \frac{1}{T_2}\right)$$

Temperature dependence of an equilibrium constant may be examined by plotting $\ln K$ versus $1/T$. From the van't Hoff equation, it can be seen

$$\left(\frac{\partial \ln K}{\partial(1/T)}\right)_P = -\frac{\Delta H_r^o}{R}$$

---

* This is generally the case when the range of temperatures involved is not appreciable, and in the absence of any phase changes in the participating species.

The temperature dependence of an equilibrium constant can thus be determined by plotting $\ln K$ versus $1/T$:

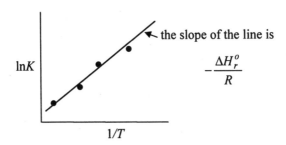

the slope of the line is
$$-\frac{\Delta H_r^o}{R}$$

As the slope of the line is positive as shown, the reaction is exothermic. In other words, $\Delta H_r^o < 0$.

Exercises

1. The equilibrium constant for the ammonia synthesis reaction

$$\tfrac{1}{2}N_2(g) + \tfrac{3}{2}H_2(g) = NH_3(g)$$

is 775 at 25°C based on 1 atm ideal gas standard states. The enthalpy change associated with the reaction, or simply the heat of reaction, ΔH_r^o, is -45.9kJ. Assuming that ΔH_r^o is independent of temperature, estimate the equilibrium constant for the reaction at 45°C.

2. Equilibrium of the system containing MnO(s), Mn$_3$O$_4$(s) and O$_2$(g) was examined. Both MnO and Mn$_3$O$_4$ were pure and stable states. It was found that equilibrium partial pressures of oxygen were

$$P_{O_2} = 1.4 \times 10^{-6} \text{ atm} \quad \text{at } 1{,}000°C$$
$$P_{O_2} = 2.8 \times 10^{-5} \text{ atm} \quad \text{at } 1{,}100°C$$

Calculate the heat of reaction ΔH_r^o of the reaction

$$3MnO(s) + \tfrac{1}{2}O_2(g) = Mn_3O_4(s)$$

Assume that ΔH_r^o is independent of T.

3.1.4. Effect of Pressure on Equilibrium Constant

The equilibrium constant depends on the value of the standard free energy change of a reaction:

$$K = \exp\left(-\frac{\Delta G_r^o}{RT}\right)$$

Recall that in the equation ΔG_r^o is the free energy change when all the reactants and products are at their respective standard states in which P_{total} = 1atm. Therefore K is independent of pressure. This conclusion does not necessarily mean that the equilibrium composition is independent of pressure. Consider the reaction

$$X_2(g) = 2X(g) \qquad K = \frac{P_X^2}{P_{X_2}}$$

K is independent of pressure. This does not mean that the individual partial pressures are independent of the total pressure, but the ratio P_X^2 / P_{X_2} is independent of the total pressure. Recall that $P_i = N_i P_{total}$. Substitution of this equation into the equation for K yields

$$\boxed{K = \frac{P_X^2}{P_{X_2}} = \left(\frac{N_X^2}{N_{X_2}}\right) P_{total}}$$

If the total pressure, P_{total}, changes, values of the individual mole fractions change in such a way that the ratio cancels the change of P_{total}.

$$\text{Let } K_C = \left(\frac{N_X^2}{N_{X_2}}\right)$$

$$\boxed{K = K_C P_{total}}$$

Independent of pressure   Dependent on pressure

For the general reaction:

$$aA(g) + bB(g) = mM(g) + nN(g)$$

$$\boxed{K = \frac{P_M^m P_N^n}{P_A^a P_B^b} = \left(\frac{N_M^m N_N^n}{N_A^a N_B^b}\right) P^{(m+n-a-b)}}$$

$$\text{Let } K_C = \frac{N_M^m N_N^n}{N_A^a N_B^b}$$

$$\boxed{K = K_C P^{(m+n-a-b)}}$$

K_C is the equilibrium constant expressed in concentrations, and note that it is independent of pressure, only if $m+n-a-b = 0$ i.e., no net change in the number of moles by the reaction.

Example 1

Consider the reaction

$$A(g) = 2B(g) \qquad K = \frac{P_B^2}{P_A}$$

If the total pressure in the reactor is increased by injecting an inert gas into the reactor, would the equilibrium constant K change?

Consider the definition of partial pressure : Partial pressures of perfect gases are the pressures that each species would exert if it were alone in the system. Therefore the presence of another gas has no effect on the equilibrium constant and on the equilibrium molar concentrations (e.g., mol/cm³) of species so long as the gases are perfect.

Example 2

Consider the reaction

$$A(g) = 2B(g) \qquad K = \frac{P_B^2}{P_A}$$

If the total pressure in the reactor is increased by compression, would the equilibrium constant K change? Would the partial pressures of the individual species change?

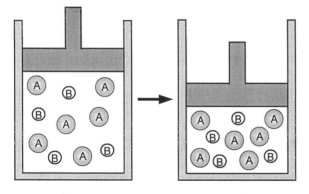

Recall that

$$K = \exp\left(-\frac{\Delta G_r^o}{RT}\right)$$

ΔG_r^o is independent of pressure and hence K is independent of pressure.

As K is independent of pressure, P_B^2 / P_A (= K) should also be independent of pressure. However, the compression, or pressure change in general, adjusts the individual partial pressures of the species in such a way that, although the partial pressure of each species changes, their ratio appearing in the equilibrium constant expression remains unchanged.

Exercises

1. The standard free energy change of the dissociation reaction

$$N_2O_4(g) = 2NO_2(g)$$

   is 4,770 J at 25°C. The system initially contains N_2O_4 only.

   1) Calculate the equilibrium partial pressure of NO_2 at 1 atm total pressure.
   2) Calculate the equilibrium partial pressure of NO_2 at 10 atm total pressure.

2. The equilibrium constant K for the reaction

$$2SO_2(g) + O_2(g) = 2SO_3(g)$$

   is 110.7 at 600°C. Assuming that the gases behave ideally, calculate the equilibrium constant K_c when the concentrations are expressed in mol per liter.
   (Gas constant $R = 0.082$ liter atm mol^{-1}K^{-1})

3.1.5. Le Chatelier's Principle

Consider the general reaction which is at equilibrium :

$$aA + bB = mM + nN$$

The reaction is then subjected to a change in conditions that affect the reaction equilibrium. This perturbation will cause the reaction to proceed toward a new equilibrium. But in which direction? Toward right or left or unaffected?
Le Chatelier's principle provides a convenient way of predicting the direction in which the reaction proceeds toward a new equilibrium state.

Le Chatelier's Principle
Perturbation of a system at equilibrium will cause the equilibrium position to change in such a way as to tend to remove the perturbation.

We do not need to evaluate the equilibrium constant K to apply Le Chatelier's principle.

(*Examples*)
- If a reaction is exothermic, the reaction will be promoted by lowering the temperature.
- If a reaction results in a change in volume, then increase in pressure will cause the reaction to proceed in the direction which results in decrease in volume.

Le Chatelier's principle provides a good guide to the effects of pressure, temperature and concentration. For a quantitative analysis, however, more rigorous treatments are required as seen in the previous sections.

Example 1

Consider the reaction which is at equilibrium

$$C(s) + CO_2(g) = 2CO(g)$$

The gases are assumed to behave ideally.

1. If the equilibrium is disturbed by adding some additional CO(g) into the reactor, in which reaction will the reaction proceed?
2. If some additional solid carbon is added in the reactor, what would happen?
3. If the total pressure in the reactor is increased by compression, in which direction will the reaction proceed?
4. The reaction is endothermic as written. In which direction will an increase in the temperature shift the reaction?

---

1. Because the concentration of CO is increased, the reaction proceeds in the direction which results in the consumption of CO(g). : To the left
2. The concentration of a solid is independent of the amount of the solid present so that there is no shift in the reaction equilibrium. : Unaffected.
3. An increase in the total pressure will shift the reaction equilibrium towards the side with the smaller number of moles of gas. : To the left.
4. An increase in temperature favours the absorption of heat (endothermic) and thus shifts the reaction equilibrium to the right.

Exercises

1. Consider the reaction at equilibrium :

$$A(s) + 2B(g) = M(s) + N(g) \quad \Delta H_r^o < 0$$

Determine the direction of the reaction for each of the following changes of the thermodynamic conditions :

1) Increase in temperature
2) Decrease in pressure
3) Increase in the concentration of B
4) Increase in the concentration of N.

3.1.6. Alternative Standard States

Recapitulation

$a_i = \dfrac{f_i}{f_i^o} \cong \dfrac{P_i}{P_i^o}$ The choice of a standard state is arbitrary, and the activity is always unity at the standard state chosen.

The activity of a component in a solution is essentially a relative quantity. From the definition of activity it follows that the numerical value of the activity of a particular component is dependent on the choice of the standard state. There is no fundamental reason for preferring one standard state over another. Convenience dictates the choice of the standard state. Up to now, we have chosen *the pure state as the standard state*. That is, a pure component in its stable state of existence at the specified temperature and 1atm pressure is chosen as the standard state. This particular choice is known as the *Raoultian standard state*.

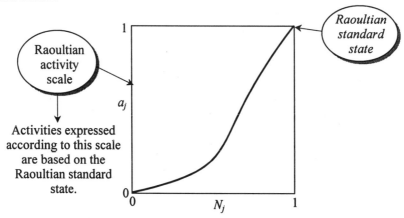

The Raoultian standard state is quite satisfactory in dealing with many solution systems. But there are some inconvenience and limitations associated with this standard state :

- If the pure component exists in a physical state which is different from that of the solution at the temperature of interest (e.g., pure oxygen is a gas, but it is liquid when dissolved in water.), how can the pressure terms in the definition of activity be determined?
- With the Raoultian standard state, it is found not infrequently that the activity of a solute in a dilute solution is very small.

To resolve these problems, we now define a new standard state called the *Henrian standard state* which originates from Henry's law.

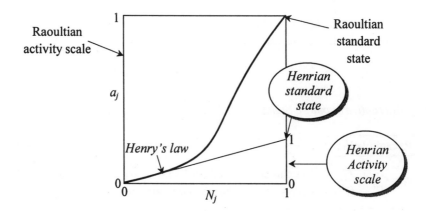

The Henrian standard state is a hypothetical, non-physical state for component j. suppose that we are interest in the composition marked x in the following figure :

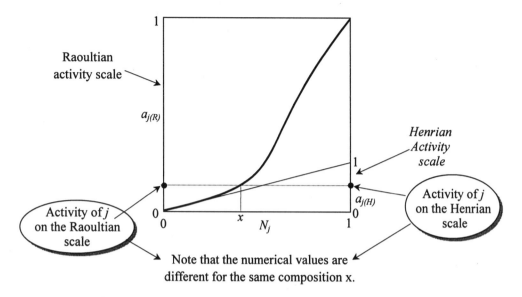

Note that the numerical values are different for the same composition x.

Two important points
1. If j obeys Henry's law, the value of the activity on the Henrian scale is numerically equal to the mole fraction of j.
2. The numerical value of the activity of j at the Henrian standard state is 1 on the Henrian activity scale, but γ_j^o on the Raoultian activity scale.

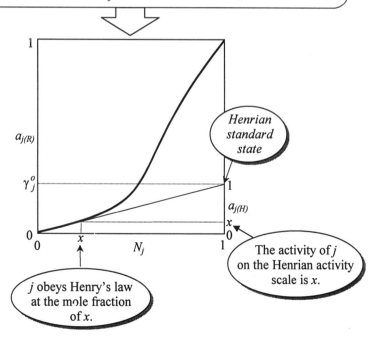

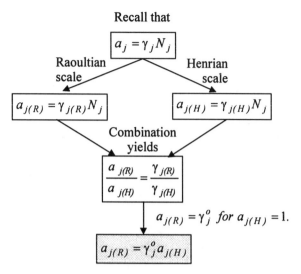

This equation relates the activity on the Henrian scale to the activity on the Raoultian scale. The Henrian standard state is sometimes called the *infinitely dilute solution standard state* because it is mostly used for dilute solutions.

The concentration of solutions is frequently given in terms of weight percent. From a practical point of view, therefore, it is more convenient to use the concentration scale expressed in weight percent (wt% j) rather than in mole fraction (N_j). We thus define a new standard state called the *1 wt% standard state*.

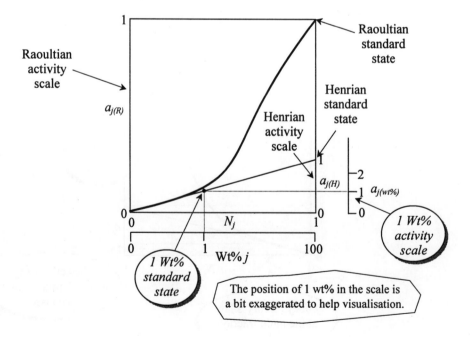

Note that both the Henrian standard state and 1wt% standard state are based on Henry's law. The difference is that the former is at $N_j = 1$ on the Henry's law line whereas the latter is at wt% $j = 1$. Note also that in dilute solutions in which j obeys Henry's law *the value of $a_{j(wt\%)}$ is numerically the same as (wt% j)*.

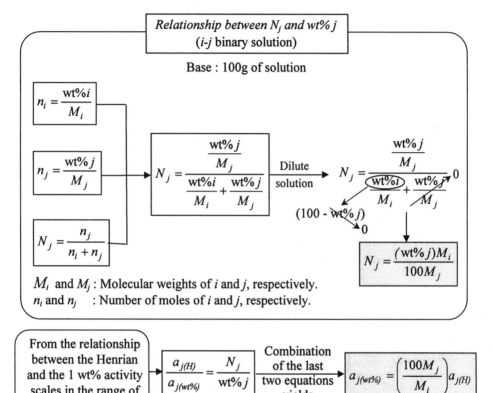

Note that the 1 wt% standard state is real if the solution obeys Henry's law up to 1 wt%, otherwise it is hypothetical.

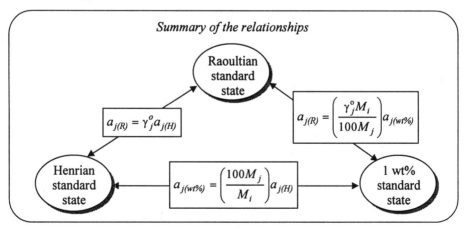

Recall that the partial molar free energy or chemical potential of a component in a solution is independent of the standard state chosen. In other words, it is an absolute thermodynamic property of the component in the solution.

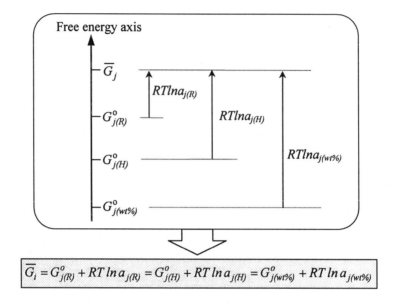

$$\overline{G}_i = G^o_{j(R)} + RT \ln a_{j(R)} = G^o_{j(H)} + RT \ln a_{j(H)} = G^o_{j(wt\%)} + RT \ln a_{j(wt\%)}$$

In thermodynamic considerations of A-B binary solution, if the standard state of B is changed from the Raoultian standard state to the Henrian standard state, the standard molar free energy changes accordingly.

$$B_{(Raoultian\ standard\ state)} \rightarrow B_{(Henrian\ standard\ state)}$$

$$\Delta G^o_{B(R \to H)} = G^o_{B(H)} - G^o_{B(R)}$$

$$\overline{G}_B = G^o_{B(R)} + RT\ln a_{B(R)}$$
$$= G^o_{B(H)} + RT\ln a_{B(H)}$$

$$a_{j(R)} = \gamma^o_j a_{j(H)}$$

$$\Delta G^o_{B(R \to H)} = RT\ln \gamma^o_B$$

In a similar way we may find the free energy change associated with the change in the standard state from the Raoultian to the 1 wt% standard state.

$$\Delta G_{B(R \to wt\%)} = RT\ln\left(\frac{\gamma^o_B M_A}{100 M_B}\right)$$

Consider the heterogeneous reaction

$$aA(l) + bB(g) = mM(s)$$

The standard free energy change at temperature T for the reaction is ΔG_r^o when Raoultian standard states, i.e., pure liquid A, gaseous B and pure solid M at 1atm are used.

$$aA\ (l, Raoultian) + bB(g, Raoultian) = mM(s, Raoultian) \qquad \Delta G_{r,1}^o$$

It is sometimes more convenient to use alternative standard states for the species involved in the reaction. When the standard state of liquid A is changed from Raoultian to Henrian standard state, the free energy change of the reaction

$$aA\ (l, Henrian) + bB(g, Raoultian) = mM(s, Raoultian) \qquad \Delta G_{r,2}^o$$

can be obtained as follows :

(1) $\quad aA\ (l, R) + bB(g, R) = mM(s, R) \qquad \Delta G_{r,1}^o$

(2) $\quad A\ (l, R) = A\ (l, H) \qquad \Delta G_{A(R \to H)}^o = RT\ln\gamma_A^o$

(3) $\quad aA\ (l, R) = aA\ (l, H) \qquad a\Delta G_{A(R \to H)}^o = aRT\ln\gamma_A^o$

(1) - (3) $aA\ (l, H) + bB(g, R) = mM(s, R) \quad \boxed{\Delta G_{r,2}^o = \Delta G_{r,1}^o - aRT\ln\gamma_A^o}$

Similarly, if the standard state of A is changed from Raoultian to 1 wt% standard state,

$$aA\ (l,\ 1\ wt\%) + bB(g, Raoultian) = mM(s, Raoultian)\ \Delta G_{r,3}^o$$

$$\boxed{\Delta G_{r,3}^o = \Delta G_{r,1}^o - aRT\ln\left(\frac{\gamma_A^o M_X}{100 M_A}\right)}$$

where M_A and M_X are molecular weights of A and solvent X in the A-X binary solution.

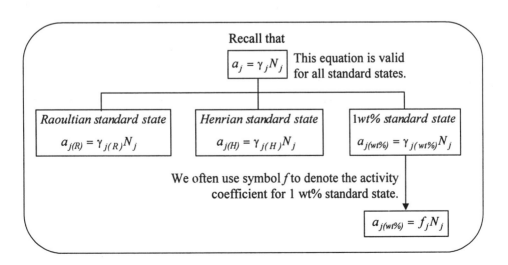

As both the Henrian and 1 wt% standard states are based on Henry's law, for dilute solutions,

$$f_j = \gamma_{j(H)}$$

Exercises

1. The activity of silicon in a binary Fe-Si liquid alloy containing $N_{Si} = 0.02$ is 0.000022 at 1,600°C relative to the Raoultian standard state. The Henrian activity coefficient γ_{Si}^o is experimentally determined to be 0.0011.

   1) Calculate the activity of silicon in the same alloy, but relative to the Henrian standard state.
   2) Calculate the change of the standard molar free energy of silicon for the change of the standard state from Raoultian to Henrian.
   3) Calculate the activity of silicon relative to the 1 wt% standard state. The molecular weights of Fe and Si are 55.85 and 28.09, respectively.

2. A liquid Fe-Al alloy is in equilibrium at 1,600°C with a Al_2O_3-saturated slag and a gas phase containing oxygen. The partial pressure of oxygen in the gas phase is maintained at 5×10^{-14} atm. Calculate the activity of aluminium in the alloy in 1 wt% standard state. The following data are given :

   $$2Al(l) + \tfrac{3}{2}O_2(g) = Al_2O_3(s) \qquad \Delta G_r^o = -1,077,500 J$$

   for all the species at Raoultian standard states.

   $$\gamma_{Al}^o = 0.029 \quad M_{Fe} = 55.85 \quad M_{Al} = 26.98$$

3. The residual oxygen present in a copper melt can be removed by equilibrating the melt with a H_2O-H_2 gas mixture. It is desired to keep the oxygen concentration in the copper melt lower than 0.001 wt% at 1,200°C. Calculate the ratio of H_2O to H_2 in the gas mixture which is in equilibrium with 0.001 wt% oxygen in the melt. The following data are given:

   $$H_2(g) + \tfrac{1}{2}O_2(g) = H_2O(g) \qquad \Delta G_1^o = -247,400 + 55.85T, J$$
   $$\tfrac{1}{2}O_2(g) = O(wt\%) \qquad \Delta G_2^o = -85,350 + 18.54T, J$$
   $$\log f_O = -0.158[wt\%O]$$

3.1.7. Interaction Coefficients

Much of the discussion thus far has been concerned with binary solutions. Most practical systems, however, are more complex and consist of several components. It is now in order to examine the thermodynamics of solutions which contain several dilute solutes. The activity of B in dilute solution with respect to the Henrian standard state is given by

$$a_{B(H)} = f_B N_B$$

where f_B is the activity coefficient of B on the Henrian activity scale.
First, suppose that we have the A-B binary solution, and let f_B^B denote the activity coefficient of B in the A-B binary solution in which B is the only solute. Then

$$f_B = f_B^B$$

Next, consider the A-B-C binary solution. Holding the concentration of B constant, we add a small amount of C. C will then influence the interaction between A and B in the solution, and hence alter the activity coefficient of B.

$$f_B = f_B^B f_B^C$$

where f_B^C is the effect of C on the activity coefficient of B.
For the A-B-C-D quaternary solution,

$$f_B = f_B^B f_B^C f_B^D$$

In this relationship it is assumed that addition of D in the solution does not give any influence on f_B^C and vice versa. This condition is generally satisfactory in most practical systems.
For a multicomponent system in general,

$$f_B = f_B^B f_B^C f_B^D \cdots$$

Taking logarithms,

$$\ln f_B = \ln f_B^B + \ln f_B^C + \ln f_B^D + \cdots$$

Since $\ln f_B$ is some function of the mole fractions of B, C, D, ..., the Taylor-series expansion yields

$$\ln f_B = N_B \left(\frac{\partial \ln f_B}{\partial N_B}\right)_{N_B=0} + N_C \left(\frac{\partial \ln f_B}{\partial N_C}\right)_{N_C=0} + N_D \left(\frac{\partial \ln f_B}{\partial N_D}\right)_{N_D=0} + \cdots$$

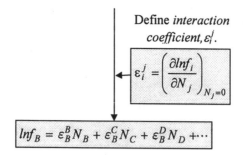

$$\ln f_B = \varepsilon_B^B N_B + \varepsilon_B^C N_C + \varepsilon_B^D N_D + \cdots$$

From a practical point of view it is often more convenient to use weight percent for expressing the concentrations in conjunction with the 1 wt% standard state and logarithms to the base ten.

$$\log f_B = e_B^B (wt\%B) + e_B^C (wt\%C) + e_B^D (wt\%D) + \cdots$$

The last two equations offer an important means in calculating the activity coefficient of a dilute solute in multicomponent solutions.

- These equations are valid only for dilute solution because of approximations taken in the Taylor-series expansion :

$$\ln f_B = \ln f_B^o + \left[N_B \left(\frac{\partial \ln f_B}{\partial N_B} \right)_{N_B=0} + N_C \left(\frac{\partial \ln f_B}{\partial N_C} \right)_{N_C=0} + N_D \left(\frac{\partial \ln f_B}{\partial N_D} \right)_{N_D=0} + \cdots \right]$$
$$+ \left[\frac{1}{2} N_B^2 \left(\frac{\partial^2 \ln f_B}{\partial N_B^2} \right)_{N_B=0} + N_B N_C \left(\frac{\partial^2 \ln f_B}{\partial N_B \partial N_C} \right)_{N_B=N_C=0} + \cdots \right] + \cdots$$

Approximations
- As the Henrian standard state is chosen, $N_B \to 0$, $f_B = f_B^o = 1$.
- For dilute solutions, the second and higher order terms are negligible.

- The following relations exist between interaction coefficients :

$$\varepsilon_A^B = \varepsilon_B^A$$

$$e_A^B = \frac{M_A}{M_B} e_B^A$$

$$e_A^B = \frac{M_S}{230.3 M_B} \varepsilon_A^B$$

where
A, B : solutes
S : solvent

Exercises

1. Oxygen dissolved in molten steels is reduced by using deoxidizing elements like Al and Si. In a steel refining process at 1,600°C, aluminium content in the melt was found to be 0.01 wt% by sample analysis. Assuming that oxygen is in equilibrium with Al in the melt and pure $Al_2O_3(s)$ which is the oxidation product, calculate the wt% of oxygen dissolved in the melt. The following data are given:

$$2Al(l) + \tfrac{3}{2}O_2(g) = Al_2O_3(s) \quad \Delta G_1^o = -1{,}682{,}900 + 323.24T, J$$

$$Al(l) = Al(l, wt\%) \quad \Delta G_{2(R \to wt\%)}^o = -63{,}180 - 27.91T, J$$

$$\tfrac{1}{2}O_2(g, 1atm) = O(l, wt\%) \quad \Delta G_{3(R \to wt\%)}^o = -117{,}150 - 2.89T, J$$

$$e_{Al}^{Al} = 0.048 \quad e_{Al}^{O} = -6.6 \quad e_{O}^{Al} = -3.9 \quad e_{O}^{O} = -0.20$$

2. A liquid copper containing dissolved oxygen and sulfur is in equilibrium with the gaseous phase consisting of N_2, O_2, SO, SO_2 and SO_3 at 1,206°C. Calculate the SO_2 partial pressure which is in equilibrium with the melt containing 0.02 wt%S and 0.1 wt%O. The following data are available:

$$\tfrac{1}{2}S_2(g) + O_2(g) = SO_2(g) \quad \Delta G_1^o = -361{,}670 + 72.68T, J$$

$$\tfrac{1}{2}O_2(g) = O(l, wt\%) \quad \Delta G_2^o = -85{,}350 + 18.54T, J$$

$$\tfrac{1}{2}S_2(g) = S(l, wt\%) \quad \Delta G_3^o = -119{,}660 + 25.23T, J$$

$$e_S^O = -0.33 \quad e_S^S = -0.19 \quad e_O^O = -0.16 \quad e_O^S = -0.16$$

3.1.8. Ellingham Diagram

The standard free energy of formation (ΔG_f^o) of a compound varies with temperature. The variation with temperature is usually presented by means of a table or some simple equations like

$$\Delta G_f^o = A + BT \ln T + CT, \quad \text{or}$$

$$\Delta G_f^o = A + BT$$

Ellingham presented the variation of ΔG_f^o with temperature in a graphical form in that ΔG_f^o was plotted against temperature. He found that relationships in general were *linear* over temperature ranges in which no change in physical state occurred. The relations could well be represented by means of the simple equation:

$$\boxed{\Delta G_f^o = A + BT}$$

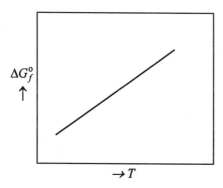

In plotting ΔG_f^o -T diagrams Ellingham made use of $\Delta \tilde{G}_f^o$ which is the standard free energy of formation of the compound, not per mole of the compound, but per mole of the gaseous element consumed. For instance,

$$2Cr(s) + \frac{3}{2}O_2(g) = Cr_2O_3(s) \quad \Delta G_f^o$$

$$\frac{4}{3}Cr(s) + O_2(g) = \frac{2}{3}Cr_2O_3(s) \quad \Delta \tilde{G}_f^o$$

Consider a general reaction of formation :

$$M + O_2(g) = MO_2 \quad \Delta \tilde{G}_f^o = A + BT$$

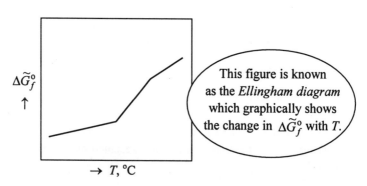

This figure is known as the *Ellingham diagram* which graphically shows the change in $\Delta \tilde{G}_f^o$ with T.

Recall that

$$\boxed{\Delta \tilde{G}_f^o = - RT \ln K}$$

$$K = \frac{a_{MO_2}}{a_M P_{O_2}} \quad \begin{array}{l} a_M = a_{MO_2} = 1 \\ \text{when } M \text{ and } MO_2 \text{ are pure.} \end{array}$$

$$\boxed{\Delta \tilde{G}_f^o = RT \ln P_{O_2}}$$

The above equation relates $\Delta \tilde{G}_f^o$ to the oxygen pressure in equilibrium with $M(s)$ and $MO_2(s)$ at temperature T. Each discontinuity point in the diagram indicates the phase change of a species involved in the formation reaction. For instance,

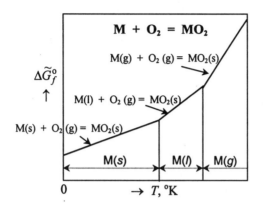

Recall that

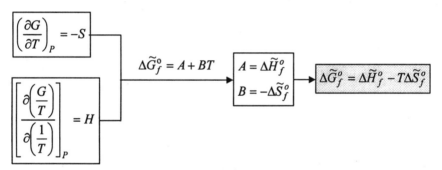

Thus the slope of the line in the Ellingham diagram is $-\Delta \tilde{S}_f^o$ and the intercept of the line at 0 K is $\Delta \tilde{H}_f^o$. Simplifying the notations in the diagram, we have

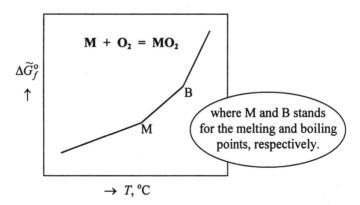

where M and B stands for the melting and boiling points, respectively.

As we may add as many formation reactions as we want, this method of presentation provides a large amount of thermodynamic data and shows the relative stability of compounds for given conditions.

The Ellingham diagram for some oxides is given below :

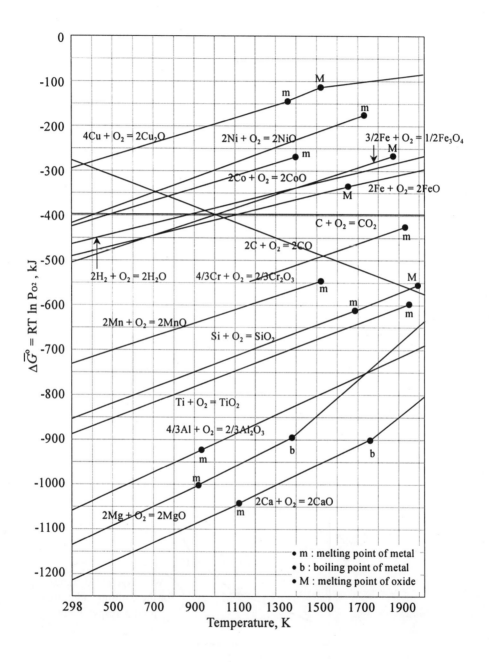

3.2. Phase Equilibria

3.2.1. Phase Rule

Suppose that we have one mole of ideal gas A.

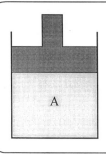

- Are we free to vary temperature? Yes.
- Are we free to vary both temperature and pressure independently? Yes.
- Can we then vary temperature, pressure and volume independently? No.

We must specify only two variables, i.e., (P,V), (P,T) or (V,T). The third variable is uniquely determined by the equation of state. In other words, the equilibrium condition of the system is fully defined by specifying two variables.
we may say that we have *two variables* under our control. We then say that the number of *degrees of freedom* = 2.

What if the gas phase is the mixture of the ideal gases A and B?

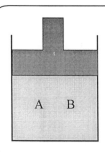

- Is the number of degrees of freedom still the same? In other words, can we fully define the thermodynamic state of the system by specifying two variables, say, T and P? No.

Specifying T and P will determine V, but the composition of the gas phase is left undetermined until the concentration of either A or B is fixed.

That is, even after fixing T and P we can freely vary the concentration of either A or B (but not both because $N_A + N_B = 1$), and hence have *one additional degree of freedom*. Thus, the number of degrees of freedom is three in this case.

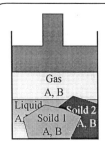

- What if a liquid phase coexists with the gas phase?
- What if a solid phase coexists as well?
- Would it be thermodynamically feasible to have two solid phases together with a liquid phase and a gaseous phase in equilibrium?

Are there systematic methods available to answer these questions?

J.W. Gibbs developed the thermodynamic methods for the characterisation of equilibrium states of heterogeneous systems involving any number of substances. Before deriving a thermodynamic method for the characterisation of equilibrium states of heterogeneous systems, we define two important terms: i.e., *phase* and *component*.

Phase

Phase is defined as a physically distinct, homogeneous and mechanically separable part of a system.

(Examples)
- The *ice – water – steam* system at 0°C has three distinct phases: solid(ice), liquid(water) and gas(steam).
- *Vapours* and *gases*, either pure or mixed, constitute a single phase because the component gases are miscible.
- *Solutions* (liquid and solid) are single phases.
- *Immiscible liquids* constitute separate phases (e.g., liquid slag and metal in a furnace)

Component

The number of components of a system is the minimum number of composition variables that must be specified in order to completely define the composition of each phase in the system.
The number of components is not necessarily the same as the number of elements, species or compounds present in the system, but given by

$$c = s - r$$

where c = number of components,
s = number of chemically distinct constituents,
r = number of algebraic relationships among the composition variables

(Examples)
In the *nitrogen-hydrogen-ammonia* system
- A non-reactive mixture of $N_2(g)$, $H_2(g)$ and $NH_3(g)$ at a low temperature

$$s = 3, r = 0 \rightarrow c = 3$$

- A mixture of $N_2(g)$, $H_2(g)$ and $H_2(g)$ at a high temperature where the following equilibrium is established:

$$N_2(g) + 3H_2(g) = 2NH_3(g) \qquad K = \frac{P_{NH_3}^2}{P_{N_2} P_{H_2}^3}$$

Thus, $s = 3, r = 1 \rightarrow c = 2$

Each independent chemical reaction at equilibrium gives rise to a restrictive condition.

- If the above mixture was initially obtained by heating $NH_3(g)$, there exists an additional restrictive condition :

$$P_{H_2} = 3P_{N_2}$$
Thus, $s = 3$, $r = 2 \rightarrow c = 1$

Note that each stoichiometric relation or constraint gives rise to a restrictive condition.

We now consider more on the equality of chemical potentials at equilibrium.

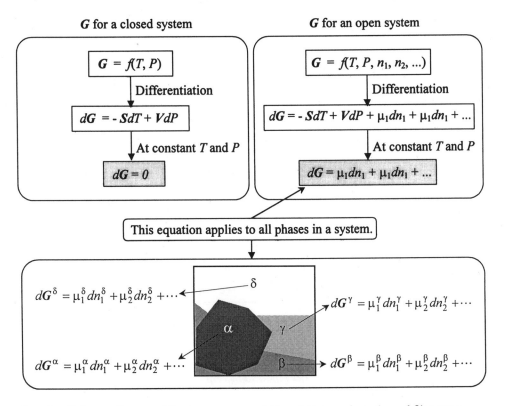

For simplicity we first consider two component (1 and 2)-two phase (α and β) system.

$$dG = dG^\alpha + dG^\beta = \mu_1^\alpha dn_1^\alpha + \mu_2^\alpha dn_2^\alpha + \mu_1^\beta dn_1^\beta + \mu_2^\beta dn_2^\beta$$

From mass balance
$$dn_1^\alpha + dn_1^\beta = 0 \text{ and } dn_2^\alpha + dn_2^\beta = 0$$

$$dG = (\mu_1^\alpha - \mu_1^\beta)dn_1^\alpha + (\mu_2^\alpha - \mu_2^\beta)dn_2^\alpha = 0 \text{ at equilibrium.}$$

Since dn_1^α and dn_2^α in the equation are arbitrary and hence not zero,

$$\mu_1^\alpha = \mu_1^\beta \text{ and } \mu_2^\alpha = \mu_2^\beta$$

At equilibrium, therefore, the chemical potential of each component is constant throughout the entire system.

We are now ready to derive an equation known as the *phase rule* which applies to both homogeneous and heterogeneous systems at equilibrium. Consider a system composed of c components that are distributed between p phases.

Number of variables	*Number of restricting equations*
There are $(c-1)$ component variables in each phase, because by knowing the concentrations (e.g., mole fractions) of $(c-1)$ components the last one can be found from $N_1 + N_2 + ... + N_c = 1$	The chemical potential of component i is constant throughout the system: $$\mu_i^\alpha = \mu_i^\beta = \mu_i^\gamma = \cdots = \mu_i^\delta$$ Hence the number of independent equations for each component is $(p-1)$.
Since there are p phases, the total number of component variables of the system is $p(c-1)$.	Since there are c components in the system, the total number of restricting equations is $c(p-1)$.
There are two additional variables, temperature and pressure.	
Thus the total number of variables is $p(c-1) + 2$	

Therefore, the number of undetermined variables is given by,

$$\boxed{\text{Number of undetermined variables} = p(c-1) + 2 - c(p-1) = c - p + 2}$$

This is called the *number of degrees of freedom* and denoted by the symbol f.

$$\boxed{f = c - p + 2}$$

This equation is called the *Gibbs phase rule* or simply the *phase rule*.

- The phase rule offers a simple means of determining the minimum number of intensive variables that have to be specified in order to unambiguously determine the thermodynamic state of the system.
- The application of the phase rule does not require a knowledge of the actual constituents of a phase
- The phase rule applies only to systems which are in equilibrium.

Example 1

Suppose we have a system which is composed of water and steam in equilibrium.
1) Can we choose at will the equilibrium temperature of the system?
2) We add ice to the system and want to have all three phases in equilibrium. Can we choose at will the equilibrium pressure of the system?

1) $c = 1$ (H$_2$O), and $p = 2$ (gas, liquid). Thus, from the phase rule, $f = 1$.
   The significance of the value $f = 1$ is that we can choose either the equilibrium temperature or the equilibrium pressure, but not both.

2) $c = 1$ (H$_2$O), and $p = 3$ (gas, liquid, solid). Thus $f = 0$
   The significance of the value $f = 0$ is that we are not allowed to freely choose either T or P. In other words the system can only exist at a unique temperature and a unique pressure. This unique point is called the *triple point*, which is discussed in Section 3.2.2. in more detail.

Example 2

Consider the substance M which has two allotropes α and β. Is it possible that four phases (solid α, solid β, liquid M and vapour of M) coexist in equilibrium?

$c = 1$ (M), $p = 4$ (α, β, liquid and vapour) → $f = -1$
The significance of a negative value of f is that the system is not capable of having all the phases enumerated coexisting in equilibrium.

Example 3

The term *phase* signifies a state of matter that is uniform throughout, not only in chemical composition, but also in physical state. Determine the number of phases in each of the systems specified in the following :

1) Ice chipped into small pieces
2) An alloy of two metals which are immiscible
3) An alloy of two metals which are miscible
4) A gas mixture composed of N$_2$, O$_2$ and CO
5) A liquid solution of A and B in equilibrium with their respective pure solids
6) A liquid solution of A and B in equilibrium with a liquid solution of their oxides

---

1) 1   2) 2   3) 1   4) 1   5) 3   6) 2

Exercises

1. Shown below is a one-component phase diagram. There exist 4 different phases, namely, α, β, γ and δ.
   1) Calculate the number of degrees of freedom in the area A.
   2) Calculate the number of degrees of freedom on the line B.
   3) Calculate the number of degrees of freedom at the point C.

4) An experimenter has reported that a new phase ε was found to coexist in equilibrium with β, γ and δ phases at a particular set of conditions. What is your opinion about this report?

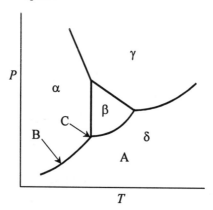

2. Consider a system in which the following equilibrium occurs :

$$CaCO_3(s) = CaO(s) + CO_2(g)$$

Calculate the number of degrees of freedom.

3. Solid zinc oxide is reduced by solid carbon at a high temperature. The system is in equilibrium and found to contain the following species :

$$ZnO(s), C(s), Zn(g), CO(g), CO_2(g)$$

1) Calculate the number of degrees of freedom.
2) Calculate the number of degrees of freedom if the system is initially prepared from $ZnO(s)$ and $C(s)$.

4. Solid solutions consisting of GaAs and InAs can be produced by equilibrating a gas mixture composed of H_2, HCl, $InCl$, $GaCl$ and As_4. The composition of the solid solution is determined by the thermodynamic conditions of the system. In order to produce GaAs-InAs solid solution with a particular composition, how many intensive thermodynamic variables need to be fixed?

3.2.2. Phase Transformations

The intensive properties of a system include temperature, pressure and the chemical potentials (or partial molar free energies) of the various species present.

- If there is a difference in *temperature*, the transfer of energy occurs as *heat*: Temperature is a measure of the tendency of heat to leave the system.

- If there is a difference in *pressure*, the transfer of energy occurs as *PV work*: Pressure is a measure of the tendency towards mechanical work

- If there is a difference in *chemical potential*, the transfer of energy occurs as *transfer of matter*. Chemical potential is a measure of the tendency of the species to leave the phase, to react or to spread throughout the phase through chemical reaction, diffusion, etc.

A substance with a higher chemical potential has a spontaneous tendency to move to a state with lower chemical potential. Why do solid substances melt upon heating? Why do liquid substances vaporise rather than solidify upon heating? Why do phase transitions occur? The chemical potential provides the key to these questions. Consider a one-component system :

$$\boxed{\left(\frac{\partial G}{\partial T}\right)_P = -S} \quad \text{or} \quad \boxed{\left(\frac{\partial \mu}{\partial T}\right)_P = -S} \quad (G = \mu)$$

This equation shows that, because entropy is always positive, the chemical potential of a pure substance decreases as the temperature is increased. As $S_{(g)} > S_{(l)} > S_{(s)}$, the slope of the plot of μ versus T is steeper for the vapour than for the liquid, and steeper for the liquid than for the solid.

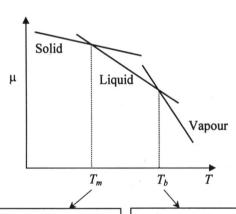

Below this temperature, $\mu_{(s)} < \mu_{(l)}$ and hence the solid phase is more stable. *Above* this temperature, on the other hand, the liquid is more stable. At T_m, the chemical potential of the solid and the liquid are the same and hence both phases coexist. T_m : Melting temperature	*Below* this temperature, $\mu_{(l)} < \mu_{(g)}$ and hence the liquid phase is more stable. *Above* this temperature, on the other hand, the gas is more stable. At T_b, the chemical potential of the liquid and the gas are the same and hence both phases coexist. T_b : Boiling temperature

A *phase transition*, the spontaneous conversion from one phase to another, occurs at a characteristic temperature for a given pressure. This characteristic temperature is called the *transition temperature*.

Next, examine the effect of pressure on the chemical potential:

$$\left(\frac{\partial \mu}{\partial P}\right)_T = V$$

As V is always positive, an increase in pressure increases the chemical potential of any pure substance. For most substances, $V_{(l)} > V_{(s)}$ and hence, from the equation, an increase in pressure increases the chemical potential of the liquid more than that of the solid.

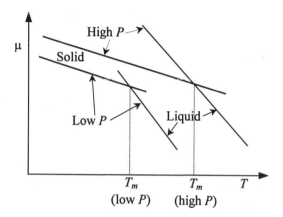

Thus an increase in pressure results in an increase in the melting temperature. If $V_{(l)} < V_{(s)}$, however, an increase in pressure effects a decrease in T_m. (e.g., $V_{water} < V_{ice}$)
It would be useful to combine the effects of temperature and pressure on phase transition of a substance in a same diagram.

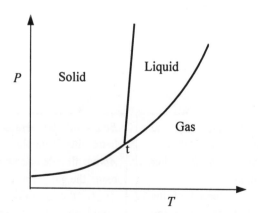

This figure is known as the phase diagram of a substance. It shows the thermodynamically stable phases at different pressures and temperatures. The lines separating phases are called as *phase boundaries* at which two phases coexist in equilibrium. The point "t" is the *triple point* at which three phases coexist in equilibrium.

Now, a question arises, "Is there a way to quantitatively describe the phase boundaries in terms of P and T?" The phase rule predicts the existence of the phase boundaries, but does not give any clue on the shape (slope) of the boundaries. To answer the above question, we make use of the fact that at equilibrium the chemical potential of a substance is the same in all phases present.

Consider the phases α and β which are in equilibrium.

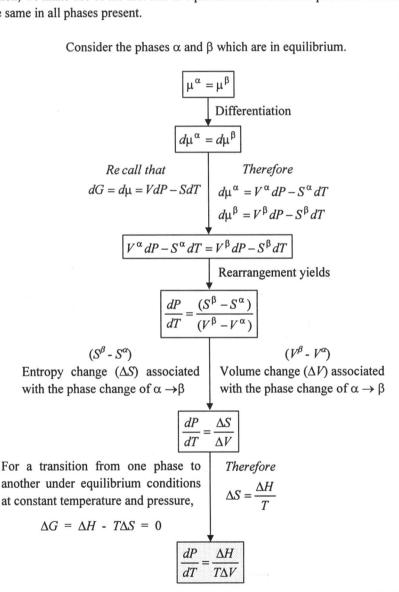

This equation is known as the *Clapeyron equation*. The Clapeyron equation is limited to equilibria involving phases of fixed composition (e.g., one-component system) because μ has been assumed independent of composition :

$$d\mu = VdP - SdT + \mu_1 dn_1 + \mu_2 dn_2 + \cdots$$

For the solid-liquid phase boundary,

$$\boxed{\frac{dP}{dT} = \frac{\Delta H_f}{T\Delta V_f}}$$

where ΔH_f is the heat of fusion, and greater than zero, ΔV_f is the volume change upon melting, and greater than zero in most cases and always small. Therefore dP/dT is positive and very large. This indicates that the slope of the solid-liquid phase boundary is positive and steep in most cases.

Integrating the Clapeyron equation assuming that ΔH_f and ΔV_f are constant,

$$\boxed{P_2 - P_1 = \frac{\Delta H_f}{\Delta V_f} \ln\left(\frac{T_2}{T_1}\right) \cong \frac{\Delta H_f}{T_1 \Delta V_f}(T_2 - T_1)}$$

For the liquid-vapour phase boundary,

$$\boxed{\frac{dP}{dT} = \frac{\Delta H_v}{T\Delta V_v}}$$

ΔH_v : Heat of vaporisation > 0 ΔV_v : Volume change upon vaporisation
$\qquad = V_{(g)} - V_{(l)} \cong V_{(g)} = RT/P$

$$\boxed{\frac{d\ln P}{dT} = \frac{\Delta H_v}{RT^2}}$$

This equation is known as the *Clausius-Clapeyron equation*. Integration yields,

$$\boxed{\ln\left(\frac{P_2}{P_1}\right) = -\frac{\Delta H_v}{R}\left(\frac{1}{T_2} - \frac{1}{T_1}\right)}$$

Using this equation ΔH_v can be estimated with a knowledge of the equilibrium vapour pressure of a liquid at two different temperatures. For the solid-vapour phase boundary (sublimation), an analogous equation is obtained by replacing ΔH_v with the heat of sublimation ΔH_s.

Example 1

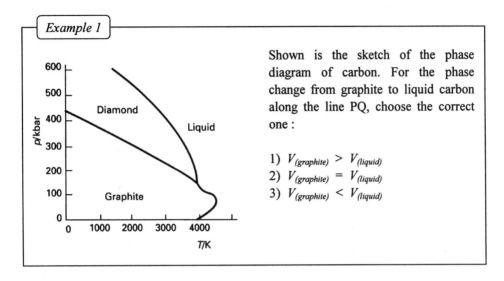

Shown is the sketch of the phase diagram of carbon. For the phase change from graphite to liquid carbon along the line PQ, choose the correct one :

1) $V_{(graphite)} > V_{(liquid)}$
2) $V_{(graphite)} = V_{(liquid)}$
3) $V_{(graphite)} < V_{(liquid)}$

From Clapeyron equation,

$$\frac{dP}{dT} = \frac{\Delta H_f}{T\Delta V_f} > 0 \text{ (from the slope)}$$

$\Delta H_f > 0$ (Fusion is endothermic.)

$$\Delta V_f = V_{liquid} - V_{graphite} > 0$$

Example 2

Substance A in the condensed phase is in equilibrium with its own vapour at temperature T. A is the only species in the gas phase. Now an inert gas is introduced into the system so that the total pressure rises from P^o to P'. Assuming that the inert gas behaves ideally with A and does not dissolve in the condensed phase, choose the correct one from the following :

1. The vapour pressure of A is independent of the total pressure of the system.
2. The vapour pressure of A is affected by the total pressure, but the change in the vapour pressure is generally negligibly small.
3. The vapour pressure of A is affected to a large extent by the total pressure.

From the knowledge that A in the condensed phase is in equilibrium with A in the gaseous phase,

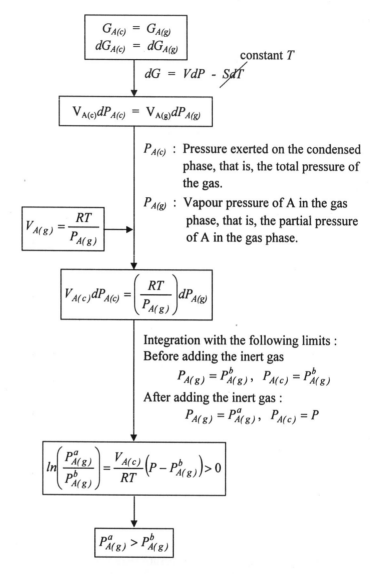

However, $V_{A(c)}/RT$ in the equation is small and hence the difference between $P_{A(g)}^a$ and $P_{A(g)}^b$ is negligibly small.

Exercises

1. A system contains ice at 0°C and 1 atm. Calculate the change in the chemical potential of ice associated with the increase in the pressure from 1 atm to 2 atm. The density of ice is 0.915 g/cm³. Calculate the change in the chemical potential of water associated with the increase in the pressure from 1 atm to 2 atm. Ice is now in

equilibrium with water at 0°C and 1 atm. If the pressure in the system is increase to 2 atm, will ice melt, freeze or remain unchanged?

2. Consider the phase transformation

$$HgS(s, red) = HgS(s, black) \quad \Delta G_r^o = 4{,}180 - 5.44T \, J$$

1) Determine the enthalpy change of the reaction.
2) Determine the transition temperature at $P = 1$ atm.
3) Determine the pressure at which both HgS(s, red) and HgS(s, black) coexist in equilibrium at 500°C.

$$M_{HgS} = 232.7 \text{ g/mol}, \quad \rho_{HgS(red)} = 8.1 \text{ g/cm}^3, \quad \rho_{HgS(black)} = 7.7 \text{ g/cm}^3$$

4) Calculate dP/dT at 495°C using Clapeyron equation.

3. The vapor pressure of liquid iron is given by the equation

$$\log P_{Fe} = \frac{-19{,}710}{T} - 1.27 \log T + 13.27 \quad \text{(torr)}$$

Calculate the standard heat of vaporization at 1,600°C

4. The vapor pressures of solid and liquid zinc are given by

$$\ln P_{Zn(s)} = \frac{-15{,}780}{T} - 0.755 \ln T + 19.3 \quad \text{(atm)}$$

$$\ln P_{Zn(l)} = \frac{-15{,}250}{T} - 2.255 \ln T + 21.3 \quad \text{(atm)}$$

Calculate the temperature at which solid, liquid and gaseous zinc coexist in equilibrium (triple point).

5. The equilibrium vapor pressures of solid and liquid NH_3 are given by

$$\ln P_{NH_3(s)} = 23.03 - \frac{3{,}754}{T} \quad \text{(torr)}$$

$$\ln P_{NH_3(l)} = 19.49 - \frac{3{,}063}{T} \quad \text{(torr)}$$

Calculate the heat of fusion (ΔH_f) of NH_3.

3.2.3. Phase Equilibria and Free Energies

When a liquid solution is cooled slowly, temperature will eventually reach the *liquidus* point, and a solid phase will begin to separate from the liquid solution.

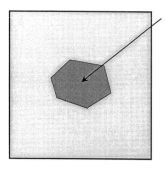

This solid phase could be a pure component, a solid solution or a compound. Now a question arises as to in which direction the system will proceed with cooling.
Precipitating
a pure component? a solid solution?
a compound? or something else?
The answer is
To the state at which the free energy of the system is minimum under the given thermodynamic conditions.

Therefore phase changes can be predicted from thermodynamic information on the free energy-composition-temperature relationship.

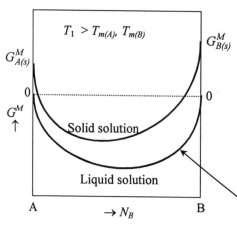

Suppose that the temperature of the A-B solution is sufficiently high so that

$$T_1 > T_{m(A)}, T_{m(B)}$$

and hence liquid is stable for both A and B. Therefore the natural choice of the standard state for A and B will be the pure liquid A and B.

$$\boxed{G^M_{A(l)} = 0} \quad \boxed{G^M_{B(l)} = 0}$$

The free energy of mixing (G^M) will change with composition.

$$\boxed{G^M = N_A G^M_A + N_B G^M_B}$$

$$G^M_i = RT \ln a_i$$

$$\boxed{G^M = RT(N_A \ln a_A + N_B \ln a_B)}$$

As the temperature T_1 is higher than the melting points of both A and B, $T_{m(A)}$ and $T_{m(B)}$, pure solid A and B will be unstable at this temperature

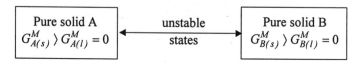

The free energy of mixing (G^M) of the solid solution, if it existed indeed, will vary with composition as shown in the figure.

At the temperature T_1, therefore, liquid solutions have free energies lower than solid solutions, and hence are stable over the entire range of composition.

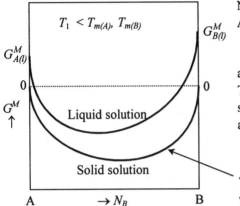

Next, suppose that the temperature of the A-B solution is sufficiently low so that

$$T_2 < T_{m(A)}, T_{m(B)}$$

and hence solid is stable for both A and B. Therefore the natural choice of the standard state for A and B will be the pure solid A and B.

$$\boxed{G^M_{A(s)} = 0} \quad \boxed{G^M_{B(s)} = 0}$$

The free energy of mixing (G^M) will change with composition.

As the temperature T_1 is lower than the melting points of both A and B, $T_{m(A)}$ and $T_{m(B)}$, pure liquid A and B will be unstable at this temperature

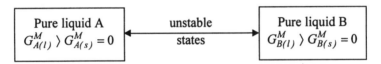

The free energy of mixing (G^M) of the liquid solution, if it existed, will vary with composition as shown in the figure.

At the temperature T_2, therefore, solid solutions are stable over the entire range of composition.

If the change in the free energy of mixing (G^M) of a mixture with composition is concave downward at constant temperature and pressure, why is homogeneous solution is stable over the entire range of composition?

Consider a mixture of A and B with an average composition of "a" shown in the following figure. If this mixture forms a homogeneous solution (α), the free energy of mixing of the solution is given by G^M_α. Is there any way to lower the free energy of mixing below this value? If the mixture forms two separate phases, β and γ, with the composition of b and c, respectively, the molar free energies of mixing are G^M_β and G^M_γ, respectively. As the average composition of the system (sum of β and γ phases) must be the same as the initial mixture (a), the proportion of each phase is given by

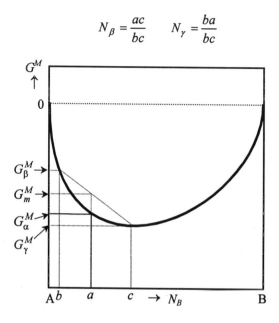

The average free energy of mixing of the mixture of β and γ phases is thus G_m^M as shown in the figure. It is clear that the phase separation has resulted in an increase in the free energy of mixing ($G_m^M > G_\alpha^M$) and hence there is no way of lowering the free energy of mixing below G_α^M.

Now we have a question as to the shape of the free energy curve: Why is the curve of G^M versus N_i concave downward?

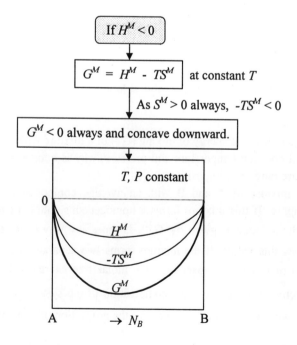

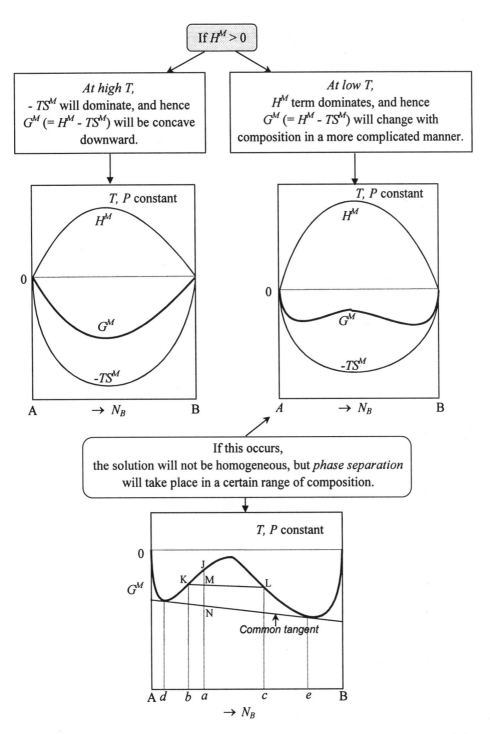

Consider a solution of composition "a" at constant temperature and pressure. If the solution does not dissociate, but maintains homogeneity, G^M of the solution is represented by "J" in the figure. It can be seen from the figure that it is possible to lower

G^M below J by dissociating into two separate coexisting solutions. For instance, if the solution dissociates into two solutions of compositions "b" and "c", respectively,

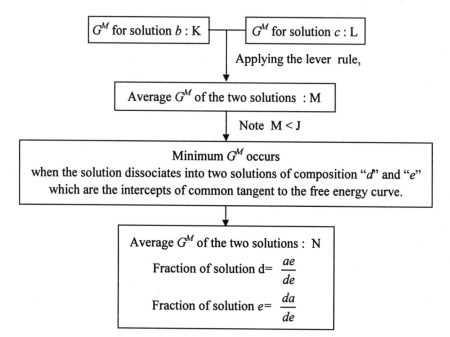

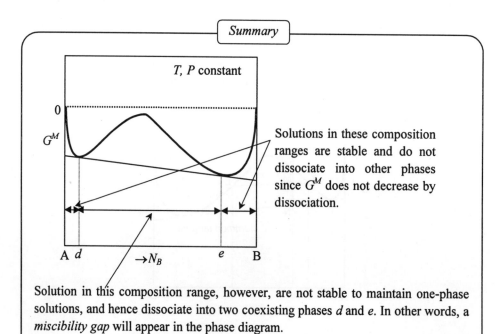

Solutions in these composition ranges are stable and do not dissociate into other phases since G^M does not decrease by dissociation.

Solution in this composition range, however, are not stable to maintain one-phase solutions, and hence dissociate into two coexisting phases d and e. In other words, a *miscibility gap* will appear in the phase diagram.

Miscibility gap

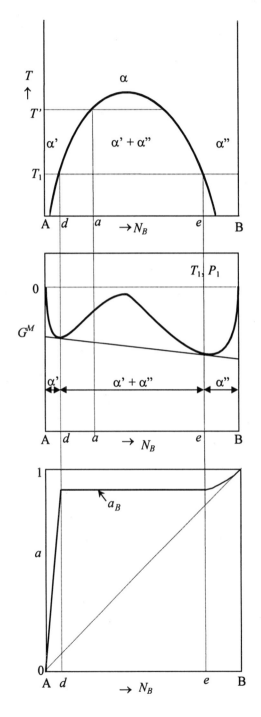

Suppose we have a solution of composition "a" at a sufficiently high temperature.

When the solution is slowly cooled, it maintains homogeneous α phase until the temperature reaches T' at which it begins to dissociate into phase α' and α''.

As the temperature is further decreased, the compositions of α' and α'' are changed along the gap boundary curve. At the temperature T_1

composition of α' : d
composition of α'' : e

$$\text{fraction of } \alpha' = \frac{ae}{de}$$

$$\text{fraction of } \alpha'' = \frac{da}{de}$$

When the average composition is changed, the proportions of α' and α'' are changed accordingly, but the compositions of α' and α'' stay at d and e, respectively, as long as the average composition lies between d and e.

Now, we examine the change in the activity of B, a_B, with the change in the overall composition. Recall that

$$G_B^M = RT \ln a_B$$

Since G_B^M is constant between d and e (common tangent), a_B is also constant in this composition range. This is obvious, because, although the average composition changes, the compositions of the individual phases (α' and α'') remain unchanged.

More discussions on the miscibility gap are given in Chapter 4.

We now consider an A-B binary solution at temperature T_2 which is above the melting point of A ($T_{m,A}$), but below the melting point of B ($T_{m,B}$).

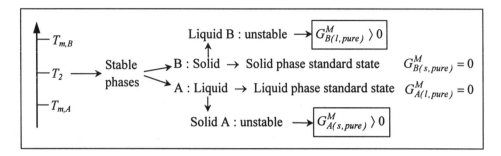

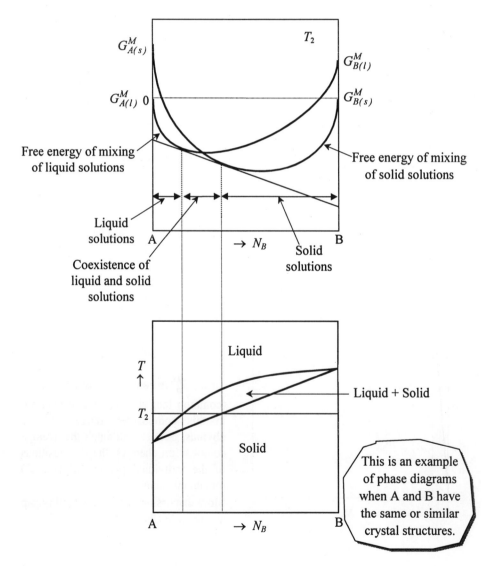

This is an example of phase diagrams when A and B have the same or similar crystal structures.

Complete solid solubility requires that components have the same crystal structure, similar atomic size, electronegativity and valency. If any of these conditions are not met, a miscibility gap will occur in the solid state.

Consider a system which consists of components A and B that have differing crystal structure and has the phase diagram as shown below :

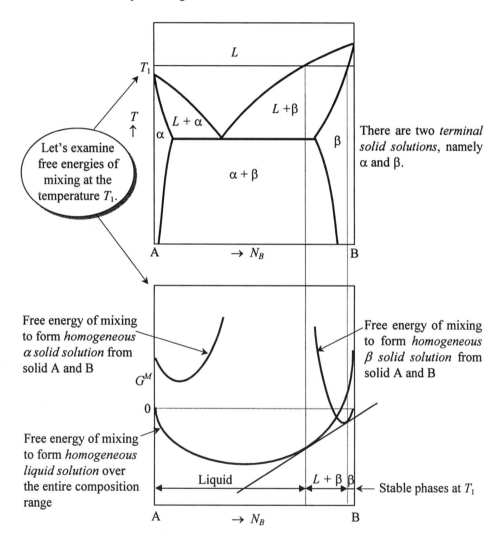

There are two *terminal solid solutions*, namely α and β.

Let's examine free energies of mixing at the temperature T_1.

Free energy of mixing to form *homogeneous* α *solid solution* from solid A and B

Free energy of mixing to form *homogeneous* β *solid solution* from solid A and B

Free energy of mixing to form *homogeneous liquid solution* over the entire composition range

Stable phases at T_1

Example 1

Prove that addition of an infinitesimal amount of a solute to a pure substance always results in decrease in free energy of the system.

The free energy of mixing of the solution, or the relative integral molar free energy, G^M, for a A-B binary solutions given by

$$G^M = N_A G_A^M + N_B G_B^M$$

$$G_i^M = RT \ln a_i, \quad a_i = \gamma_i N_i$$

$$G^M = RT(N_A \ln N_A + N_B \ln N_B) + RT(N_A \ln \gamma_A + N_B \ln \gamma_B)$$

$N_A + N_B = 1$ and differentiation

$$\left(\frac{\partial G^M}{\partial N_B}\right)_{P,T} = RT \ln\left(\frac{N_B}{1-N_B}\right) - RT \ln \gamma_A + RT \ln \gamma_B + RTN_A \frac{\partial \ln \gamma_A}{\partial N_B} + RTN_B \frac{\partial \ln \gamma_B}{\partial N_B}$$

When $N_B \to 0$, $\dfrac{N_B}{1-N_B} \to 0$

and hence

$$\ln\left(\frac{N_B}{1-N_B}\right) \to -\infty$$

Therefore,
the change in G^M with N_B at $N_B \cong 0$ is negative,
irrespective of the values of γ_A and γ_B.

Example 2

If the temperature is not much lower than the true melting point (T_m), the free energy of fusion can be estimated by the following equation :

$$\Delta G_f = \Delta H_f^o \left(1 - \frac{T}{T_m}\right)$$ where ΔH_f^o : enthalpy of fusion at T_m

Prove the above equation.

At the true melting point.

$$\Delta G_f^o = \Delta H_f^o - T_m \Delta S_f^o$$

$\Delta G_f^o = 0$

$$\Delta S_f^o = \frac{\Delta H_f^o}{T_m}$$

At the temperature T,

$$\Delta G_f = \Delta H_f - T_m \Delta S_f$$

Change in ΔH and ΔS with a small change in T is negligibly small. Thus $\Delta H_f \cong \Delta H_f^o$, $\Delta S_f \cong \Delta S_f^o$.

$$\Delta G_f = \Delta H_f^o \left(1 - \frac{T}{T_m}\right)$$

Example 3

The *critical temperature*, T_C, associated with the miscibility gap in the phase diagram shown, is the temperature below which phase separation occurs.
If A-B binary solution behaves regularly, the critical temperature is given by

$$T_C = \frac{\Omega}{2R}$$

where Ω is the interaction parameter.
Using the figure shown below, prove the above equation.

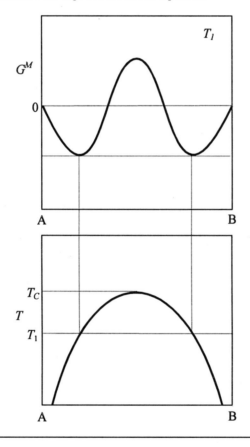

$$G^M = H^M - TS^M$$

Recall that, for the regular solution,
$H^M = N_A N_B \Omega$
$S^M = -R(N_A \ln N_A + N_B \ln N_B)$

$$G^M = \Omega N_A N_B + RT(N_A \ln N_A + N_B \ln N_B)$$

This is the equation which represents the free energy curve in the figure.
- Below T_C, the equation gives two minima and hence two inflection points as shown.
- Above T_C, the equation gives a curve concave downward.
- At T_C, the two minima and the two inflection points coincide. Mathematically,

$$\boxed{\frac{\partial G^M}{\partial N_B} = \frac{\partial^2 G^M}{\partial N_B^2}}$$

Combination of the above two equations yields

$$\boxed{T = \frac{2N_B(1-N_B)\Omega}{R}}$$

The maximum T occurs at $N_B = 0.5$. $\left(\dfrac{\partial T}{\partial N_B} = 0\right)$

$$\boxed{T_C = \frac{\Omega}{2R}}$$

Example 4

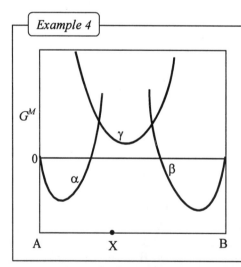

Shown is the free energy of mixing versus composition diagram of the A-B binary system at the temperature T and the pressure P. The diagram shows two *terminal phases*, α and β, and one *intermediate phase* γ. If the overall composition of the system is given by the point X shown in the diagram, find the stable equilibrium phase(s) at T and P.

The minimum value of free energy the system with the composition of X can have is found by drawing the common tangent to the free energy curves of α and β phases. For the system at the overall composition of X, the equilibrium structure is the mixture of phase α and phase β with the following proportions (refer to the figure below.):

$$\text{Fraction of } \alpha = \frac{Xb}{ab} \quad \text{Fraction of } \beta = \frac{aX}{ab}$$

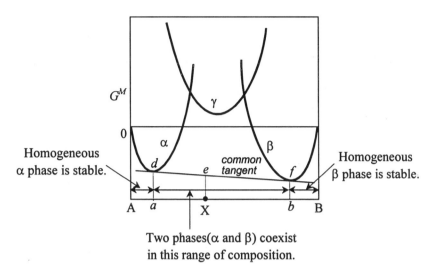

Two phases (α and β) coexist in this range of composition.

Note that γ phase does not exist as a stable phase at the present temperature and pressure. In other words, if γ phase is seen in the structure, then the system is not in the equilibrium, but a meta-stable, non-equilibrium state. However, if the thermodynamic state of the system is changed (e.g., different temperature or pressure), γ phase may exist as a stable phase in a certain composition range :

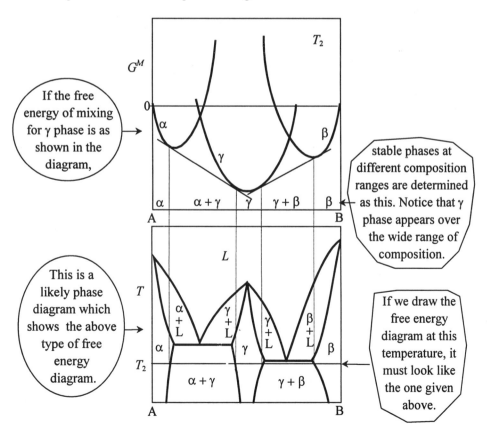

Exercises

1. The A-B binary system conforms to regular solution behaviour and the interaction parameter Ω is 17,400 J/mol. Find the critical temperature T_C below which phase separation occurs. Calculate the compositions of α' and α'' in equilibrium at 900K.

2. At 1273K, a copper-zinc alloy containing 16mol% Zn lies on the solidus, and that on the liquidus contains 20.6mol% Zn. The activity coefficient of Zn in liquid Cu-Zn alloys, relative to the *pure liquid zinc standard state*, is represented by

$$RT\ln\gamma_{Zn} = -19,246N_{Cu}^2 \quad \text{where } R = 8.314 Jmol^{-1}K^{-1}$$

Find activity of copper in the alloy of the solidus composition, relative to the *pure solid copper standard state*. The standard free energy of fusion of copper is given by

$$\Delta G_{f,Cu}^o = 6,883 - 8.745\ln T + 3.138x10^{-3}T^2 + 53.739T, \quad Jmol^{-1}$$

3. Shown below is the diagram of free energy of mixing of liquid and solid solutions of the A-B binary system at temperature T_1.
   1) Is it true that $T_1 < T_{m(A)}$ and $T_1 < T_{m(B)}$?. $T_{m(i)}$ is the melting point of i.
   2) The diagram shown occurs when the heat of mixing, H^M, is positive. Prove that
   $$H_{solid}^M \rangle H_{liquid}^M$$

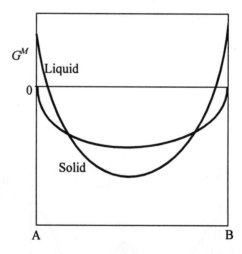

CHAPTER 4

PHASE DIAGRAMS

4.1. Unary Systems

4.1.1. Pressure-Temperature Diagrams

Phase changes are effected by three externally controllable variables. These are *pressure*, *temperature* and *composition*. In a one-component system, or *unary system*, however, the composition does not vary, but must always be unity. Therefore there are only two variables which can vary: pressure and temperature. Every possible combination of temperature and pressure can be readily represented by points on a two-dimensional diagram.

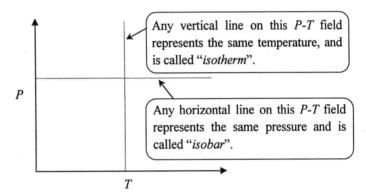

The three states or phases, namely solid, liquid and gas, can be represented by the corresponding areas in the *P-T* field :

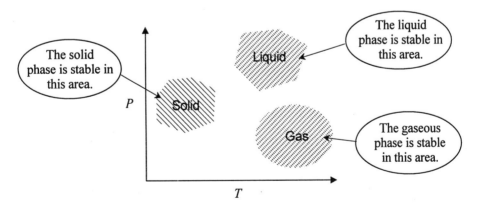

Boundaries separating these phases are called *phase boundaries.*

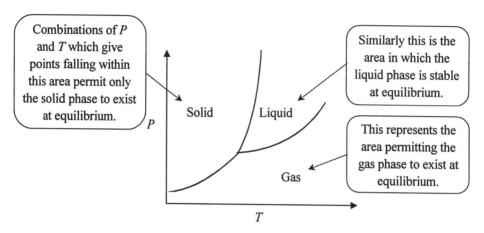

Points falling directly on the phase boundaries represent conditions which require that two neighbouring phases coexist at equilibrium :

e.g.: point a : coexistence of the solid and gas phases
point b : coexistence of the solid and liquid phases
point c : coexistence of the liquid and gas phases

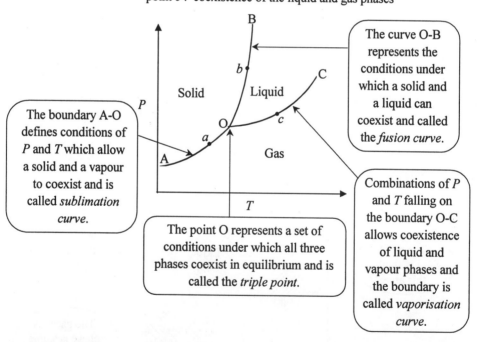

The vaporisation curve (O-C) does not extend indefinitely, but ends abruptly at point C which is called the *critical point*. Let's examine the physical significance of the critical point.

Suppose that a liquid is contained in a sealed vessel and in equilibrium with its vapour at the temperature T_1 (Fig A). When the liquid is heated, the vapour pressure increases and hence the density of the vapour phase increases, but the quantity of the liquid decreases due to vaporisation and the density of the liquid decreases due to thermal expansion (Fig B). Eventually there comes to a stage at which the density of the vapour becomes equal to that of the remaining liquid and the interface between the two phases disappears and thus the two phases are indistinguishable (Fig C).

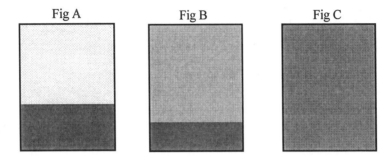

Fig A   Fig B   Fig C

The temperature at which the interface disappears is called the *critical temperature* (T_C), and the corresponding vapour pressure is called the *critical pressure* P_C. At and above T_C, therefore, the liquid phase does not exist.

Now we discuss dynamic changes which take place when a system under equilibrium is disturbed by change in temperature or pressure. Consider a system represented by point k, in which the liquid and vapour phases are in equilibrium at T_1 and P_1. If the temperature of the system is suddenly increased to T_2 while *keeping the total volume constant*, the initial equilibrium conditions are disturbed and the position of the system is now shifted to point m. This appears to put the system into a single phase (vapour phase) region. The position of m is however not the equilibrium one for the system, because the volume of the system is kept constant. What should happen in reality is that the thermal energy supplied to raise temperature will cause the liquid to vaporise and hence increase the pressure of the system to P_2 so that a new equilibrium is established at point n.

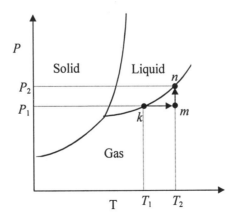

> **Example 1**
>
> The P-T diagram for the substance A is given below:

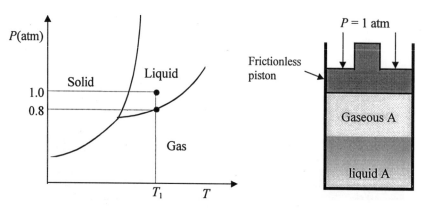

> 1) If a cylinder (system) containing pure liquid A and pure gaseous A in equilibrium with the external pressure of 1 atm at the temperature T is brought to a new condition where the temperature is T_1 and the pressure is 1 atm, what change would occur to the system?
>
> 2) If the gaseous phase in the cylinder is initially not the pure A, but a mixture of A and an inert gas, what would be the pressure of A in the gas phase at the new condition?

1) Since the external pressure of the cylinder is 1 atm, the vapor pressure of the gaseous A should be 1 atm. If conditions do not allow for the gaseous A to maintain the pressure at 1 atm, the gaseous A cannot exist under the conditions. From the P-T diagram given above, the gaseous A can exist up to 0.8 atm of its pressure at T_1, and hence the gaseous A cannot exist in equilibrium with the liquid A. Therefore all the gaseous A will liquefy at 1 atm and T_1.

2) This is not a true unary system, but a binary system of A and the inert gas. In this case, the vapor pressure of A will be adjusted so that the gaseous A at 0.8 atm exists in equilibrium with the liquid A at T_1. The partial pressure of the inert gas will be 0.2 atm to maintain the total pressure of 1 atm.

> **Example 2**
>
>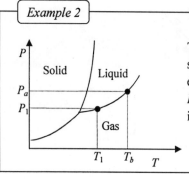
>
> The figure shown is the phase diagram of the substance A. P_1 is the vapour pressure of A in equilibrium with the liquid A at the temperature T_1. P_a indicates the ambient pressure. If the temperature is raised to T_b, what would happen to the liquid?

At the temperature T_b, the vapour pressure of A is equal to the ambient pressure. Therefore vaporisation occurs throughout the bulk of the liquid A. The condition of free vaporisation throughout the liquid is called *boiling*. Note that, in the presence of inert gas(es), a liquid does not suddenly start to form a vapour at its boiling temperature, but even at lower temperatures there is a vapour which is in equilibrium with the liquid. (See Example 1)

Example 3

A liquid sample is allowed to cool at a constant pressure, and its temperature is monitored. The results are shown in Figure A. The phase diagram of this substance is given in Figure B. Which point in the phase diagram represents the process b → c in the cooling curve?

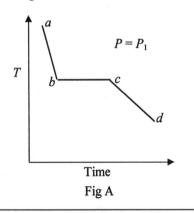

Fig A

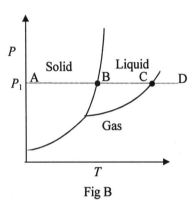

Fig B

---

2) the sample temperature will eventually hit the phase boundary at point B.

1) When a liquid sample is allowed to cool at a constant pressure P_1,

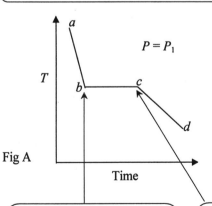

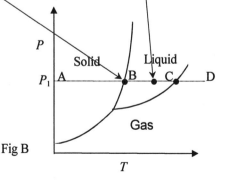

3) and then the sample begins to solidify. This is represented by the point b in the cooling curve.

4) During the phase transition from liquid to solid at the point B, heat is evolved and the cooling stops until the transition is complete at the point c in the cooling curve.

> 5) Once the transition has been complete at the point c, the sample temperature decreases again toward A from B in the phase diagram, or along the c-d in the cooling curve.

Example 4

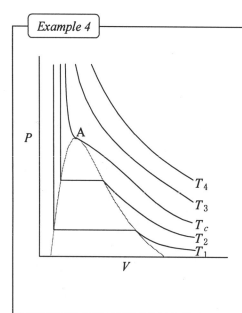

The relationship between pressure and volume is examined with a substance A. As the pressure is increased keeping the temperature at T_1, the volume of the gas decreases. When the pressure reaches a certain value, the volume suddenly reduces to the low value as the gas liquefies. A further increase in pressure causes little further reduction in volume as liquid is generally not compressible to a large extent. When the P-V relationship is examined at a number of different temperatures, one may obtain results shown in the diagram. Discuss the physical significance of the point A in the diagram.

At the point A the liquid and vapour are indistinguishable and the densities are identical. The point A is thus the critical point and the corresponding temperature T_C is the critical temperature and P_C is the critical pressure. All gases exhibit this type of behaviour, but the values of critical pressure and critical temperature vary considerably from one substance to another.

Example 5

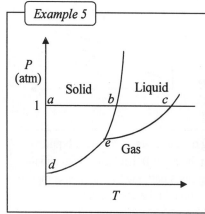

For the majority of metals the triple point lies far below atmospheric pressure and the critical point well above atmospheric pressure. The diagram given below is the P-T phase diagram of the metal M. Discuss the change of the vapour pressure when the metal is gradually heated from the solid state (a) to the temperature above the boiling point.

The behaviour of the metal during heating under a constant pressure of 1 atm can be predicted from the diagram :
- At point *a*, the solid metal is in equilibrium with the gas phase consisting of the vapour of M at the partial pressure of *d* and air (or inert gas) at the partial pressure of *d-a*, to make the total pressure of 1 atm.
- As the temperature is increased, the partial pressure (i.e., the vapour pressure) of the metal vapour increases along the curves *de* and *ec*. Note that the metal M melts at *b* during heating.
- At point *c*, the vapour pressure of the metal M becomes identical to the ambient pressure (1atm), and hence the liquid metal boils.
- Further heating will result in the metal being completely in the vapour phase.

Exercises

1. Calculate the number of degrees of freedom at the triple point.
2. Calculate the number of degrees of freedom at the critical point.
3. A metal is sealed in a completely inert and pressure-tight container. Initially the metal fills half of the container, the balance of the space being filled with an inert gas at a pressure of 1 atm. What changes will occur within the container as the system is heated. The phase diagram of the metal is similar to the that given in Example 5.

4.1.2. Allotropy

Let's look at the fundamental difference between liquids and solids :

		Liquids	Solids
General atomic array		Random arrays of atoms and molecules	Arrays of atoms or molecules in regular patterns
One component system	Atomic array	Only one type of random array is possible.	Many types of ordered arrays are possible.
	Number of phases	Only one liquid phase can exist.	A number of different solid phases may exist.

The various crystalline forms in which a solid may exist are called *allotropes* or *polymorphs*. The transformation from one allotrope to another is called *allotropic* or *polymorphic transformation*. This transformation may occur either with pressure change or with temperature change.

Suppose there are two possible solid phases, say α and β, in a given substance, and the free energies of α and β vary as shown in the following figure:

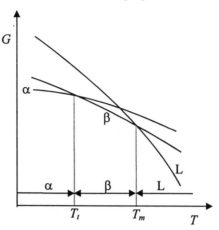

Note that

- $G_\alpha < G_\beta$ at $T < T_t$ and hence α phase is more stable,
- $G_\beta < G_\alpha$ at $T > T_t$ and hence β phase is more stable, and
- $G_{liq} < G_\beta$ at $T > T_m$ and hence the liquid phase is more stable.

The free energy curves corresponding to phase stability are then
- the curve for α in the α phase stable region,
- the curve for β in the β phase stable region and
- the curve for the liquid in the liquid phase stable region.

T_t is the temperature of allotropic transformation from α to β phase and T_m is the melting point. By combining with the effect of pressure we can construct a *P-T* diagram for a substance of interest. At a given pressure, say, the atmospheric pressure, P_a, as shown in the following diagram, as the α phase solid is heated, it transforms to the β phase at T_t, and the β phase, upon further heating, melts at T_m, and the liquid boils at T_v.

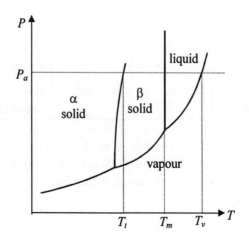

Example 1

α and β are two solid phases which are possible to appear in the substance A. At low temperatures α phase is more stable. Prove that one of the thermodynamic requirements for the appearance of allotropic transformation from α to β at constant pressure is

$$\text{Entropy of } \beta\,(S_\beta) > \text{Entropy of } \alpha\,(S_\alpha)$$

Different phases, say, α and β, in a given substance in general have different free energies and entropies at a specific temperature and pressure. First, let's look at the α phase.

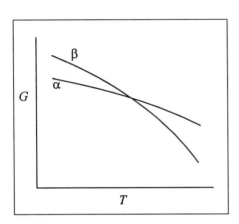

Recall that

$$\left(\frac{\partial G_i}{\partial T}\right)_P = -S_i = \text{the slope of the free energy curve}$$

Thus the slope of the curve is the negative entropy of the α phase. In order for β phase to appear at high temperatures, the free energy of β should fall more rapidly than that of α as the temperature is raised. In other words the free energy curve for β is steeper than that for α. This means that the entropy of the β phase should be larger than that of the α phase.

Example 2

The following diagram (Fig A) shows the relation between a stable phase and an unstable phase in a given set of conditions. If cooling is carried out slowly from the liquid phase, the liquid solidifies at T_m to form solid β phase. An allotropic transformation from β to α phase occurs at T_t. On slow heating the process will be reversed.
Thus the equilibrium can be expressed as

$$\alpha \leftrightarrow \beta \leftrightarrow L$$

If cooling is carried out rapidly, however, a supercooled liquid phase may appear below the temperature T_m. Further cooling may result in either further supercooling or solidifying into α phase which is metastable.
If the curve for *b* phase is located differently as shown in Fig B, can the phase β be obtained directly from the phase α?

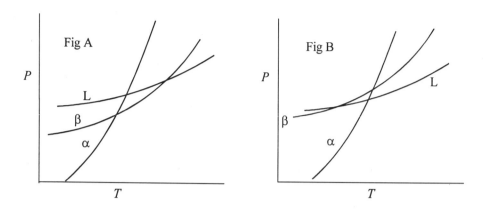

Note from Fig B that the transition point from α to β phase lies above the melting point of both α and β phases. Thus the transition from α to β cannot be observed. The only way to produce the β phase is
- to melt the α phase,
- and then to rapidly cool the melt (supercooling).

Exercises

1. The following is the phase diagram of sulphur. The stable form of sulphur at ordinary temperature is α sulphur which has a rhombic crystal structure. At what temperature will rhombic sulphur melt upon slow heating? If the sulphur is heated rapidly, at what temperature will it melt?

4.2. Binary Systems

4.2.1. Binary Liquid Systems

When two liquids are brought together, they may
- totally dissolve in one another in all proportions,
- partially dissolve in one another, or
- be completely immiscible.

Total miscibility

Consider the A-B binary liquid system in equilibrium with the vapour phase at a constant temperature. Is the composition of the vapour the same as that of the liquid? Not necessarily. Let's apply the Gibbs-Duhem equation to the liquid phase.

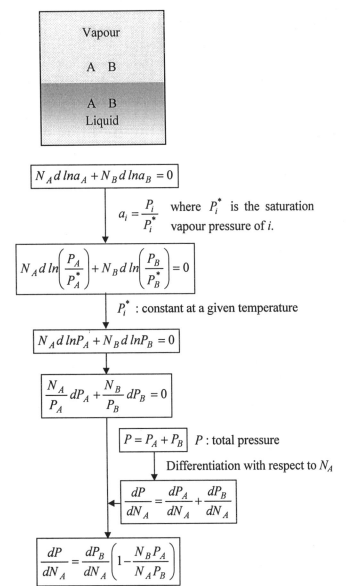

First we apply this equation to an ideal solution. Recall that an ideal solution is the one in which all components obey Raoult's law, which states that the partial pressure of a component in solution is directly proportional to its molar concentration.

Let's assume A is more volatile than B : i.e., $P_A^* > P_B^*$. Consider an arbitrary composition N_A (Refer to the following figure.). P is the total vapour pressure which is in equilibrium with the liquid solution of composition N_A. Now a question arises as to how the fraction of A in the vapour phase (P_A/P) is related to the fraction of A in the liquid phase (N_A). To derive a quantitative relationship, we need to make use of the last equation :

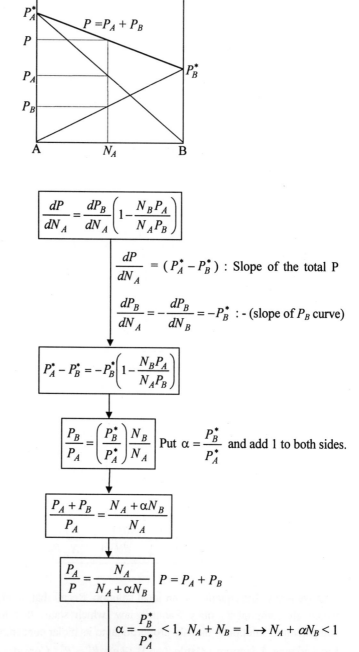

This last equation tells us that even for the ideal solution the mole fraction of a component in the vapour phase is not the same as that in the liquid phase. In the A-B binary solution of the above example in which A is more volatile than B, the vapour phase is richer in A than the liquid phase.

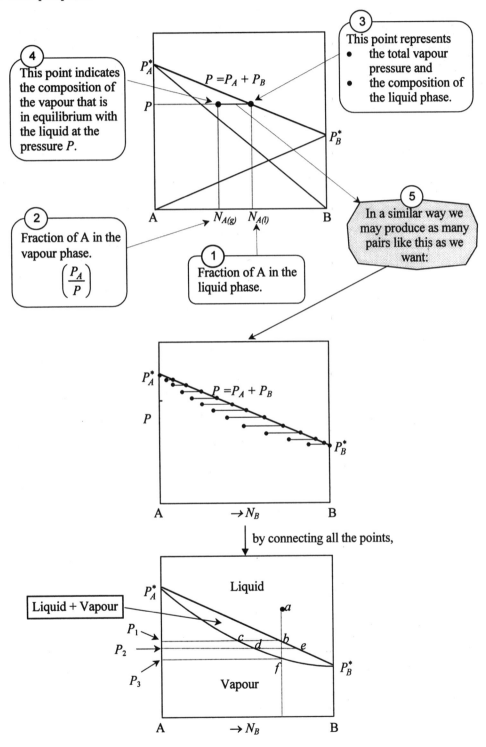

Consider lowering the pressure on a liquid solution of "a" in the above diagram.
- Until the pressure is lowered to P_1, the sample maintains a single liquid phase.
- Right at the point b, the vapour phase begins to appear, and hence the liquid coexists with its vapour of the composition c. According to the lever rule, the amount of the vapour phase is found to be negligibly small.
- Further decrease in pressure results in the change in both the composition and the relative amount of each phase. At the pressure P_2, the compositions of the vapour and the liquid are given by the points d and e, respectively. Again, the relative amounts of the liquid and vapour are determined by the lever rule.
- If the pressure is further reduced to P_3, the composition of the vapour (f) becomes the same as the overall composition. This means that at this pressure the liquid vaporises nearly completely and thus the amount of liquid is virtually zero.
- Decrease in pressure below P_3 will bring the system to the region where only vapour is present.

Next we consider a temperature-composition diagram.

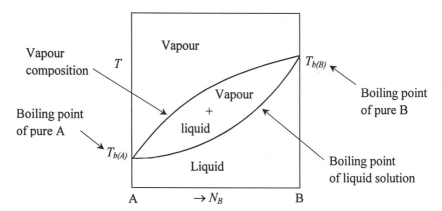

Consider the heating process of a liquid represented by the point p in the following diagram :

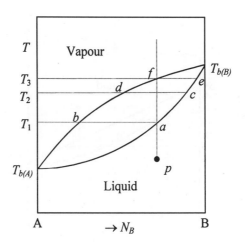

- Until the temperature of the system reaches T_1, only the liquid phase exists.
- At T_1 the liquid begins to boil. The compositions of the liquid and vapour are given by the points a and b, respectively. However, the amount of the vapour is negligibly small. Note that the vapour is richer in A than the liquid, since A is more volatile.
- On further heating, the compositions of both the liquid and the vapour are re-adjusted : the liquid composition from a toward c along the curve (*liquidus*), and the vapour from b toward d along the curve (*vaporus*).
- At T_2 the liquid of c coexists in equilibrium with the vapour of d. The proportion of each phase is determined by the lever rule.
- At T_3 the composition of the vapour (*f*) becomes equal to that of the original liquid. This means that the liquid vaporises at this temperature nearly completely and hence the liquid of the composition e is present only as a trace.
- At temperatures higher than T_3 only vapour phase can exist.

Although temperature-composition phase diagrams of many liquids are similar to the one for an ideal solution shown above, there are a number of important solutions which exhibit a marked deviation.

Recall the following equation which we have developed earlier :

$$\boxed{\frac{dP}{dN_A} = \frac{dP_B}{dN_A}\left(1 - \frac{N_B P_A}{N_A P_B}\right)}$$

When the total vapour pressure curve for a liquid solution shows a minimum or a maximum, i.e.,

$$\frac{dP}{dN_A} = 0,$$

$$\boxed{\frac{P_A}{P_B} = \frac{N_A}{N_B}}$$

It is obvious that the boiling point curve in temperature-composition diagrams will show a maximum or minimum. As

$$\frac{dP_B}{dN_A} \neq 0,$$

Therefore the composition of the vapour is the same as that of the liquid.

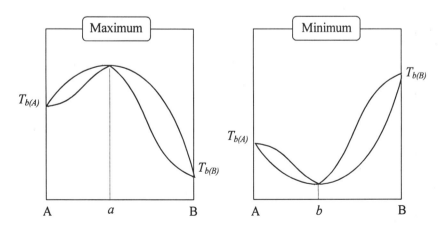

Note that at the maximum or minimum point the liquidus and vaporus curves are coincident and thus the composition of the vapor is the same as that of the liquid. In other words, evaporation of those liquid solutions denoted by *a* and *b* in the above diagrams occurs without change of composition. This type of mixture is said to form a *azeotrope*.

Liquid mixtures showing no miscibility

Certain liquids are completely immiscible in each other :

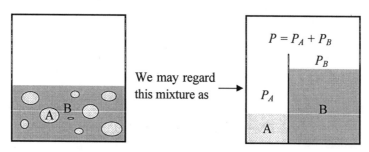

The liquid A is in equilibrium with its vapour at the vapour pressure of P_A, and the liquid B is in equilibrium with its vapour at the vapour pressure of P_B. The total pressure (P) above the liquid mixture is thus the sum of the two vapour pressures :

$$P = P_A + P_B$$

When the mixture is heated in an open container, it boils at $P = 1$ atm, not $P_A = 1$ atm or $P_B = 1$ atm. This means that any mixture of immiscible liquids will boil at a temperature below the boiling point of either component. This is the basis of *steam distillation*.

Liquid mixtures with partial miscibility

We have so far discussed two cases, namely complete solubility of liquids and complete immiscibility of liquids. But there are a number of liquid systems which show *partial miscibility*; i.e., liquids that do not mix in all proportions at all temperatures.

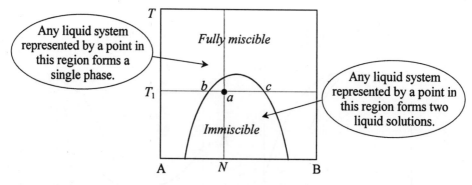

When a liquid mixture of the mean composition N is prepared at the temperature T_1, two immiscible liquid solutions will form ; one of the composition *b* and the other of the composition c. This point a in the diagram indicates merely the mean composition of the two immiscible solutions.

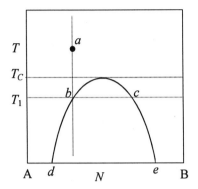

When a homogeneous liquid solution represented by the point *a* in the above diagram is cooled, the phase separation into two immiscible solutions (*b* and *c*) occurs at the temperature T_1. On further cooling compositions of both liquids change along the phase boundaries (*bd* and *ce*). T_C is the highest temperature at which phase separation can occur and is called the *critical temperature*.

Given below are diagrams which are representative of partial miscibility or phase separation :

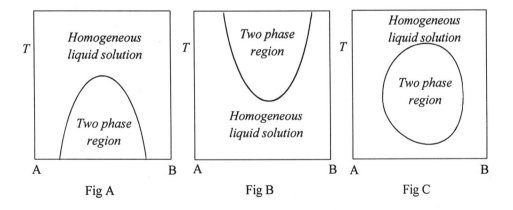

As can be seen in Fig B, some systems show a *lower critical temperature* below which they form a homogeneous liquid solution in all proportions and above which they separate into two phases. Some systems have both upper and lower critical temperatures (Fig C).

We now consider systems which include a vapour phase together with liquid mixtures with partial miscibility. The first example is a system which shows partial miscibility and a minimum boiling point azeotrope.

This example shows that the liquids become fully miscible before they boil.

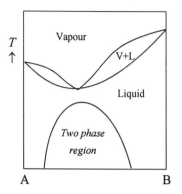

In some systems an upper critical temperature does not occur and the liquids boil before mixing is complete.

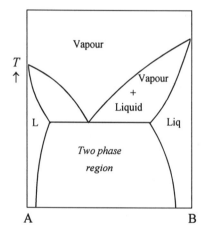

> **Example 1**
>
> A liquid of composition *a* is heated
>
>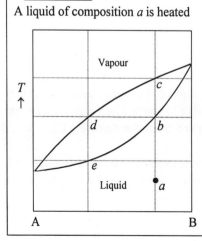
>
> 1. Find the point which represents the temperature and composition of the liquid when it begins to boil.
> 2. Find the point which represents the temperature and composition of the vapour in equilibrium with the boiling liquid.
> 3. Now the vapour of the above question is withdrawn and completely condensed and then reheated. Find the point which represents the temperature and composition of the liquid at the boiling point.

1. b,   2. d,   3. e

If the boiling and condensation cycle is repeated successively, some interesting consequences will be resulted in. Notice in the following figure that the condensed liquid becomes richer in the more volatile component A as the cycle is repeated. This process is called *fractional distillation*. Almost pure A may be obtained by repeating the cycle.

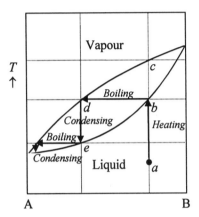

The efficiency of the fractional distillation depends very much on the shape of the phase diagram.

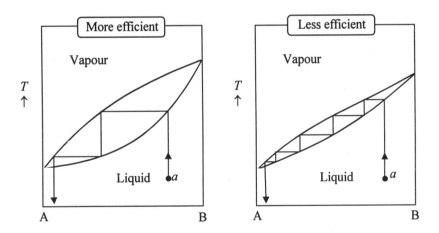

The efficiency is sometimes quantified by the number of the *theoretical plates*, the number of effective vaporisation and condensation steps that are required to achieve a condensate of given composition form a given distillate.

Example 2

Liquids A and B dissolve completely in one another in all proportions. If the change in vapour pressure $P (= P_A + P_B)$ with the fraction of A, N_A, is positive, i.e.,

$$\frac{dP}{dN_A} > 0$$

prove that the vapour phase is richer in component A that the liquid phase.

Recall the following equation :

$$\boxed{\frac{dP}{dN_A} = \frac{dP_B}{dN_A}\left(1 - \frac{N_B P_A}{N_A P_B}\right)}$$

Given

$$\frac{dP}{dN_A} > 0 \qquad \frac{dP_B}{dN_A} = -\frac{dP_B}{dN_B} < 0$$

$$\boxed{1 - \frac{N_B P_A}{N_A P_B} < 0}$$

$$\boxed{\frac{P_A}{P_B} > \frac{N_A}{N_B}}$$

Exercises

1. Liquids A and B exhibit a miscibility gap as shown in the following phase diagram. A mixture of 60 mol% of A and 40 mol% of B was prepared at 600°C. Calculate the mole fraction of the liquid rich in A.

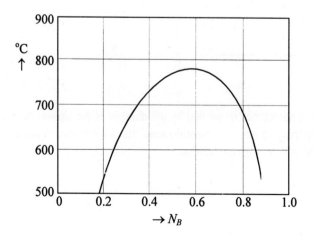

4.2.2. Binary Systems without Solid Solution

Eutectic Systems

Consider a system of two components, A and B, which are completely soluble in one another in the liquid state, but completely insoluble in one another in the solid state.
The melting point of a liquid is normally depressed if the liquid contains some other substance in solution.

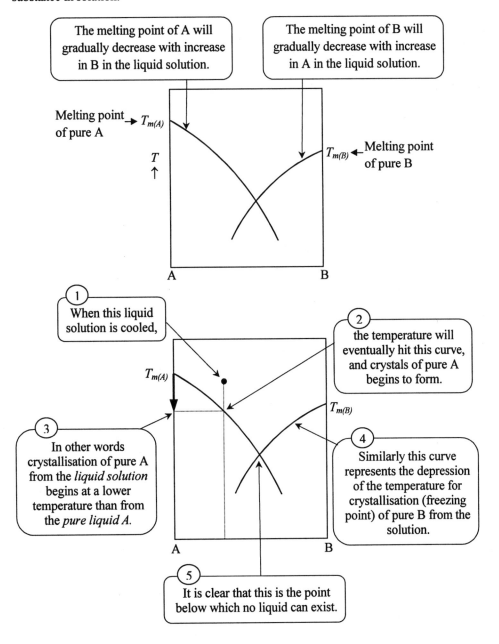

Considering all the facts discussed above, we are able to draw a diagram which shows the temperature-stable phase relationship of a binary system in which no solid solution forms. The completed *equilibrium diagram* or *phase diagram* may look like the following figure:

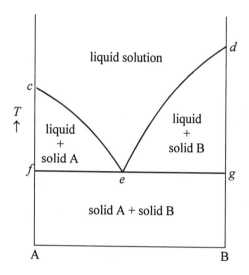

- The region above the broken curved line *ced* represents liquid solution. Any possible combination of temperature and composition which lies in this region will be completely molten. The lines *ce* and *ed* are called *liquidus*.

- The region below the horizontal line *feg* represents the mixture of solid A and B. Any possible combination of temperature and composition which lies in this region will be completely solid. The line *feg*, or more precisely, *cfegd* is called *solidus*.

- In the regions surrounded by the liquidus and solidus (*cfe* and *deg*), a solid phase coexists with a liquid phase. Consider the region *cfe*. Within this region, pure solid A can coexist with a number of different liquid solutions. Since solid A is the only solid present in this region, it is called the *primary field* of A. Similarly the region of *deg* is the primary field of B.

- The point *e* represents the lowest temperature at which a liquid solution can exist. At this point all remaining liquid solution solidifies. This point is called the *eutectic point*.

We may extract a considerable amount of information from this type of phase diagram. Let's consider the case represented by the point *p* in the following figure. We know that this point is in the primary field of A in which pure solid A coexists with liquid. When an isothermal line passing the point of interest (*p*) is constructed, the point *r* represents the composition of the liquid phase which exists in equilibrium with pure solid A represented by the point *q*. The isothermal line *qr* which connects the compositions of the two phases that coexist in equilibrium is called the *tie line*.

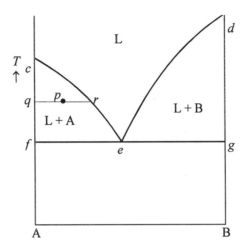

The relative amount of each phase is determined according to the lever rule :

$$\text{Fraction of solid A} = \frac{pr}{qr} \qquad \text{Fraction of liquid} = \frac{qp}{qr}$$

The microstructure at point p therefore would look like the figure given below :

We now consider the eutectic point e. If the temperature is just above the eutectic temperature, it is a single liquid phase that is present. If the temperature is just below the eutectic temperature, however, two solid phases, solid A and solid B, are present. At the eutectic point, therefore, all three phases, i.e., solid A, solid B and liquid, coexist in equilibrium. We may thus write a reaction which occurs at the eutectic point :

$$L(e) = A(s) + B(s)$$

This is called the *phase reaction* for the *eutectic reaction*. On cooling at the eutectic point the reaction will proceed to the right. On heating the reaction will proceed to the left. Solids produced by the eutectic reaction is in general a fine grain mixture of A and B.

We now examine the cooling of a composition from the liquid state to the complete solid state.

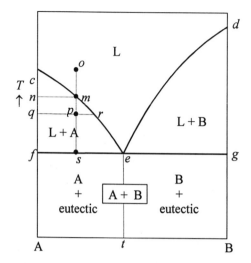

- When the liquid solution *o* is cooled, it remains in liquid until the temperature reaches the point *m*.

- At *m*, pure solid A(*n*) begins to precipitate.

- On further cooling the system enters the two phase region, and at *p* the system consists of pure solid A(*q*) and liquid solution of *r*. Note that the liquid composition has been changed from **m** to **r** along the liquidus *mr*.

- At just above the eutectic temperature, the composition of the liquid in equilibrium with solid A is *e*.

- The fraction of the liquid = fs/fe.

- On further cooling to just below the eutectic temperature, the remaining liquid which has the eutectic composition will freeze immediately according to the eutectic reaction. The solid structure will thus be the mixture of the primary phase of A and the eutectic structure which is the fine mixture of A and B.

- The solid phase region *fABg* thus can be divided into two sub-regions: *fAte* for solid A + eutectic and *etBg* for solid B + eutectic.

In a binary system two components may undergo a chemical reaction to form a compound. The compound formed may possess a definite or *congruent* melting point. If this is the case, the compound forms a separate phase and possesses a different crystal structure from those of the constituents. From many points of view a compound with a congruent melting point can be regarded as a pure substance. This type of compound will coexist in equilibrium with a liquid of identical composition. In other words the compound does not decompose below its melting point. The phase diagram of a binary system which forms a congruently-melting compound is somewhat different from the one shown above.

The phase diagram of a binary system as shown below is effectively two simple eutectic diagrams linked together. Therefore there are two eutectic points in this kind of the system. A maximum point appears on the liquidus, which is the melting point of the congruently-melting compound.

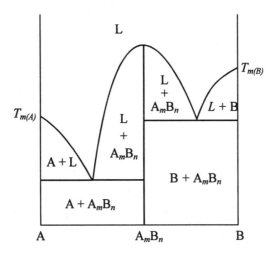

Peritectic Systems

Some compounds do not have a sharp melting temperature, but rather decompose on heating into a liquid and another solid below the liquidus temperature. When the compound A_mB_n is heated, it decomposes into solid A and liquid of p at the temperature T_P. The compound A_mB_n is stable only up to the temperature T_P. This type of compounds is called an *incongruently-melting* compound.

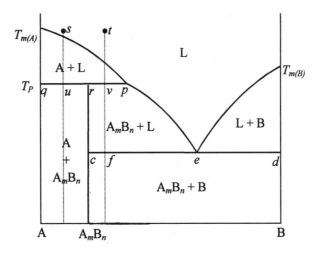

Consider the cooling of the liquid solution s.
1) Until the temperature reaches the liquidus, the solution is fully molten.
2) When the solution crosses the liquidus, primary crystals of pure A form.
3) On further cooling the crystals of A grow and the liquid becomes richer in B.
4) At a temperature just above T_P, the phases in equilibrium are pure solid A and the liquid of the composition of p. The relative amount of each phase is determined by the lever rule :

$$\text{Fraction of solid A} = \frac{up}{qp} \qquad \text{Fraction of liquid} = \frac{qu}{qp}$$

5) At the temperature T_P the crystallised A reacts with liquid and forms compound A_mB_n.

$$L(\text{at } p) + A(\text{at } q) = A_mB_n(\text{at } r)$$

6) This kind of reaction is called a *peritectic reaction* and the point p is called the *peritectic point*. The temperature T_P is called the *peritectic temperature*.
7) The peritectic reaction proceeds until all the remaining liquid is exhausted.
8) Just below the peritectic temperature, the coexistence of two solid phases, A at q and A_mB_n at r, is resulted in. Thus the microstructure of the system will consist of large primary crystals of A and small crystallites of A_mB_n.

Next, consider the cooling of the liquid solution t :
1) Until the temperature reaches the liquidus, the solution is fully molten.
2) When the solution crosses the liquidus, primary crystals of pure A form.
3) On further cooling the crystals of A grow and the liquid becomes richer in B.
4) At a temperature just above T_P, the phases in equilibrium are pure solid A and the liquid of the composition of p. The relative amount of each phase is determined by the lever rule:

$$\text{Fraction of solid A} = \frac{vp}{qp} \qquad \text{Fraction of liquid} = \frac{qv}{qp}$$

5) At the temperature T_P the crystallised A reacts with the liquid and forms compound A_mB_n.

$$L(\text{at } p) + A(\text{at } q) = A_mB_n(\text{at } r)$$

The peritectic reaction proceeds until all the primary solid A is completely exhausted. Since the quantity of B in the original composition is excessive for formation of A_mB_n alone, the crystallisation process does not end after the peritectic reaction, but continues to proceed by further cooling.

6) Just below the peritectic temperature, the coexistence of the liquid phase at p and A_mB_n at r, is resulted in.
7) On further cooling the crystals of A_mB_n grow and the liquid becomes richer in B along the liquidus *pe*.
8) At the temperature T_e, the remaining liquid (fraction =*cf/ce*) undergoes the eutectic reaction.
9) The final solid consists of large crystals of A_mB_n and small crystals of both A_mB_n and B in a eutectic matrix.

Monotectic Systems

The phase diagram given below shows a two-phase region consisting of two different liquids which are not miscible.

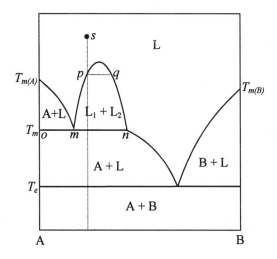

Consider the cooling of the liquid solution of s.
1) At the liquidus (p), the melt begins to separate into two liquids, L_1 (at p) and L_2 (at q).
2) As the temperature decreases, the compositions of the liquids alter along the liquidus pm and qn, respectively.
3) At the temperature T_m, the liquid phase L_1 (at m) decomposes into pure solid A (at o) and liquid L_2 (at n). This reaction may be written as

$$L_1 \text{ (at } m\text{)} = A \text{ (at } o\text{)} + L_2 \text{ (at } n\text{)}$$

This kind of reaction is called a *monotectic reaction* and the point m is referred to as the *monotectic point*.

Example 1

It is possible, for any conceivable combination of temperature and total composition, to determine by inspection of the phase diagram exactly what phases will be present at equilibrium. It is also possible to determine the exact amount of each particular phase present under any given est of conditions.

Consider a simple eutectic system shown in the following figure. When the melt s is cooled, find
1) the fraction of the primary solid phase of A just above the eutectic temperature T_e,
2) the fraction of the eutectic structure (mixture of fine crystals of A and B) just below T_e,
3) the fraction of A in the eutectic structure below T_e,
4) the fraction of A in total, i.e., the sum of A in the primary phase and the eutectic.

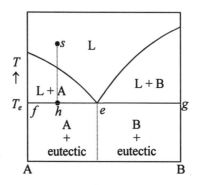

1) According to the lever rule, the fraction of A = he/fe.
2) The liquid fraction just above the eutectic temperature is converted into the eutectic structure when the temperature is lowered below the eutectic.

$$\text{Fraction of the eutectic structure} = \frac{fh}{fe}$$

3) The liquid composition at the eutectic reaction is represented by e. This liquid is transformed into solid A and solid B by the eutectic reaction.

$$\text{Fraction of A in the eutectic structure} = \frac{eg}{fg}$$

4) Fraction of the total A = hg/fg.

Example 2

Is the peritectic point in a binary system *invariant*?

Consider the Gibbs phase rule,

$$f = c - p + 2$$

$c - p$: chemical contribution
2 : temperature and pressure
However, the pressure is fixed for the phase diagram. Thus 2 → 1.

$$f = c - p + 1$$

Refer to the binary peritectic diagram in the text.
$c = 2$ (A and B)
$p = 3$ (A, liquid and A_mB_n)

$f = 0$ Invariant.

Example 3

If one of components or compounds in a binary system undergoes polymorphic transformations, there will be horizontal lines in the phase diagram separating the stable region of each polymorph. The figures given below represent two possible forms which a system can take when such transformations take place. Prove that conditions lying on these horizontal lines are univariant.

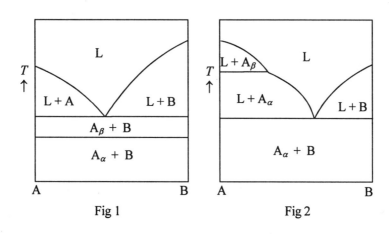

Fig 1 Fig 2

From the Gibbs phase rule,

$$f = c - p + 1$$

$c = 3$ (A_α, A_β, B)
$p = 3$ (A_α, A_β, B in Fig 1, A_α, A_β, L in Fig 2)

$f = 1$ univariant

Exercises

1. In the eutectic alloy system AB, the compositions of the three conjugate phases of the eutectic are pure A, pure B and liquid of 80% B. Assuming equilibrium solidification of an alloy composed of 40% A and 60% B at a temperature just below the eutectic temperature, calculate the percentage of the primary A. Calculate the percentage of the total A.

4.2.3. Binary Systems with Solid Solution

It is possible for solids to form a solution, i.e., *solid solution*. The concept of either liquid solution or gaseous solution is familiar and easy to conceive of. By the term solid solution, it simply means that the solute component enters and becomes a part of the crystalline solvent, without altering its basic structure. This is not limited to solids involving elements, but applies equally to solids involving compounds.

There are two kinds of solid solutions, namely, *substitutional* solid solutions and *interstitial* solid solutions. In substitutional solid solutions, the solute element occupies a position of one of the solvent elements in the solvent crystal. In interstitial solid solutions, on the other hand, the solute element occupies on of the vacant spaces between solvent elements in the solvent crystal lattice without displacing a solvent element.

Total Solid Solubility

Solid solutions with complete solid solubility, i.e., solid solubility over the entire range of the composition, are possible to form, but always of the substitutional kind. For a metallic binary solution to exhibit a complete solid solubility, for instance, both metals must have the same type of crystal structure, because it must be possible to replace, progressively, all the atoms of the initial solvent with solute atoms without causing a change in crystal structure.

For a binary system in which two components are mutually soluble in all proportions in both the liquid and solid states, the possible phase diagram shapes are as shown below:

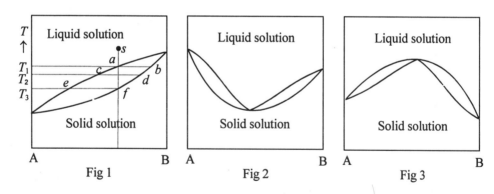

These diagrams are similar in shape to those discussed in section 4.2.1. (Liquid solution systems) and the interpretation of the diagrams is also much the same. First, consider the cooling of the liquid **s** in Fig.1.
1) Freezing of the liquid solution will commence at T_1. Crystals which begin to form at this temperature are a solid solution of the composition b, but the amount of the solid solution forming at T_1 is as a trace. Nevertheless, the liquid solution of a is in equilibrium with the solid solution b at T_1.
2) As the temperature falls, the composition of the solid solution changes following the solidus and the composition of the liquid solution changes following the liquidus. At the

temperature T_2, the liquid solution of c and the solid solution of d coexist in equilibrium. The relative amount of each phase is determined by the lever rule.
3) At T_3, solidification is complete, and further cooling will bring the system to the solid phase.

Fig 2 and 3 show the minimum and maximum melting points, respectively. The solid solution which corresponds to the minimum or maximum melting point behaves much like a pure component. It melts and freezes undergoing no changes in composition. In other words it melts congruently.

Partial Solid Solubility

In many cases, atom size, crystal structure or other factors restrict the ease with which solute atoms can be dissolved in the solvent in the solid state. Thus it is much more common to find that solids are partly soluble in one another rather than be either completely soluble or completely insoluble. The following is an example of a phase diagram for a binary system which shows *partial solid solubility* :

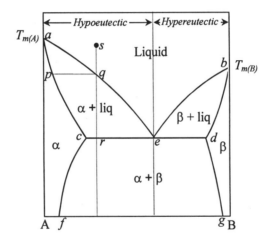

where α, β : solid solutions.
       ae, be : liquidus
       ac, cd, bd : solidus
       cf, dg : *solvus*.

Note that the solvus cf denotes the solubility limit of B in A, and the solvus dg shows the solubility limit of A in B.

Consider the cooling of the solution s.

1) At q, solidification begins. On further cooling the solid composition changes along the solidus pc, and the liquid composition along the liquidus ae.
2) When the system arrives at the eutectic temperature, the liquid left undergoes the eutectic transformation:

$$L \text{ (at } e\text{)} = \alpha \text{ (at } c\text{)} + \beta \text{ (at } d\text{)}$$

3) On completion of the eutectic reaction, the resulting structure will be the mixture of the primary α phase and the *eutectic structure* which is the mixture of α and β phase.

In a binary phase diagram it is customary that the more common component is put on the left. Those structures which occur on the left side of the eutectic composition are called *hypoeutectic*, and those on the right side are called *hypereutectic*.

If a substance is allotropic this will affect the shape of phase diagrams for systems involving the substance. Consider a system which involves two allotropic substances, A and B. The following figure shows one of the possible diagrams which involve allotropic substances. The point e in the diagram is called the *eutectoid point*, and the *eutectoid reaction* is

$$\gamma \text{ (at } e\text{)} = \alpha \text{ (at } c\text{)} + \beta \text{ (at } d\text{)}$$

Interpretation of the eutectoid phase diagram is generally the same as that of the eutectic phase diagram.

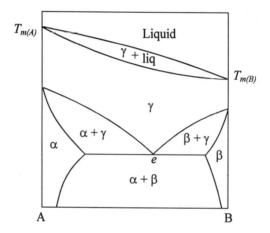

Now we consider another type of phase diagram as shown below:

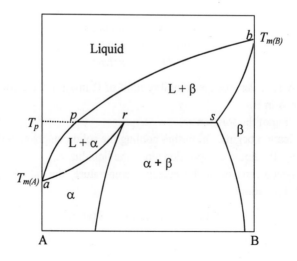

- As B in the α phase increases, the temperature at which liquid begins to form on heating rises along the solidus *ar*.

- The α phase at *r* decomposes upon heating into a liquid phase of *p* and the solid β phase at *s*. This reaction can be represented by

$$\alpha \text{ (at } r) = L \text{ (at } p) + \beta \text{ (at } s)$$

- This reaction is called the peritectic reaction, and the point *p* is known as the peritectic point and T_p the peritectic temperature.

Next we examine the cooling behaviour of several different total compositions with the figure given below.

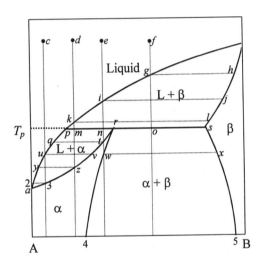

- Cooling of the liquid *c*
  1) $c \to u$: Homogeneous liquid solution
  2) At *u* : Precipitation of solid α of the composition *v*
  3) $u \to 3$: Increase in solid α phase. The compositions of the liquid and the α phase change along the *u*2 and *v*3, respectively.
  4) At 3 : Completion of solidification. The composition of α phase is given by 3.
  5) Below 3: Homogeneous α phase

- Cooling of the liquid *d*
  1) $d \to k$: Homogeneous liquid solution
  2) At *k* : Precipitation of solid β phase of the composition *l*
  3) $k \to m$: Increase in the solid β phase. The compositions of the liquid and the β phase change along *kp* and *ls*, respectively. The relative amount of each phase is determined by the lever rule.
  4) At *m* : Peritectic reaction : Portion of liquid *p* reacts with the solid β to form solid α at *r*.

$$L \text{ (at } p) + \beta \text{ (at } s) = \alpha \text{ (at } r)$$

On completion of the peritectic reaction, the system consists of the liquid p and the solid α at r.

5) $m \rightarrow z$: Increase in the α phase. The compositions of the liquid and the α phase change along py and rz, respectively.
6) At z : Completion of solidification.
7) Below z : Homogeneous α phase

- Cooling of the liquid e
  1) $e \rightarrow i$: Homogeneous liquid solution
  2) At i : Precipitation of solid β phase of the composition j
  3) $i \rightarrow n$: Increase in the solid β phase. The compositions of the liquid and the β phase change along ip and js, respectively. The relative amount of each phase is determined by the lever rule.
  4) At n : Peritectic reaction : Portion of liquid p reacts with the solid β to form solid α at r.

$$L \text{ (at } p) + \beta \text{ (at } s) = \alpha \text{ (at } r)$$

On completion of the peritectic reaction, the system consists of the liquid p and the solid α at r.

5) $n \rightarrow t$: Increase in the α phase. The compositions of the liquid and the α phase change along pq and rt, respectively.
6) At t : Completion of solidification.
7) $t \rightarrow w$: Homogeneous α phase
8) At w : Precipitation of β phase of the composition x
9) Below w: Mixture of the α and β phases. The compositions of the α and β phases change along the solvus $w4$ and $x5$, respectively. The relative amount of each phase is determined by the lever rule.

- Cooling of the liquid f
  1) $f \rightarrow g$: Homogeneous liquid solution
  2) At g : Precipitation of solid β phase of the composition h
  3) $g \rightarrow o$: Increase in the solid β phase. The compositions of the liquid and the β phase change along gp and hs, respectively. The relative amount of each phase is determined by the lever rule.
  4) At o : Peritectic reaction : All liquid p reacts with a portion of the solid β to form solid α at r.

$$L \text{ (at } p) + \beta \text{ (at } s) = \alpha \text{ (at } r)$$

On completion of the peritectic reaction, the system consists of the solid β at s and the solid α at r.

5) Below o: Mixture of the α and β phases. The compositions of the α and β phases change along the solvus $r4$ and $s5$, respectively. The relative amount of each phase is determined by the lever rule.

Given below are some other types of binary phase diagrams

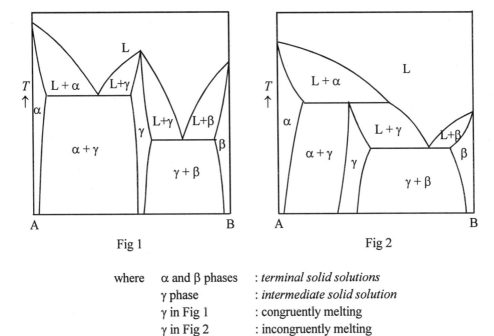

Fig 1

Fig 2

where   α and β phases   : *terminal solid solutions*
       γ phase   : *intermediate solid solution*
       γ in Fig 1   : congruently melting
       γ in Fig 2   : incongruently melting

Interpretation of these diagrams is much the same as those discussed previously.

Example 1

The Fig 1 in the following shows part of the phase diagram of the A-B binary system. When a liquid sample of the composition x was cooled to the room temperature, it was found that each crystal or grain was *richer in A toward the centre* (Fig 2). Discuss the solidification process which would enable the formation of this non-uniform concentration grain structure and determine whether the structure shown in Fig 2 is the equilibrium one.

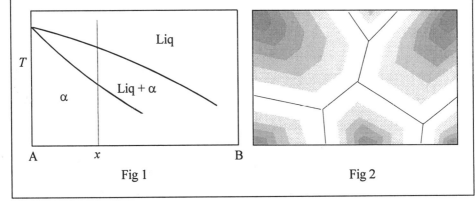

Fig 1

Fig 2

True equilibrium freezing is almost unattainable in practice, because it requires that both phases, the liquid and the solid, be homogeneous throughout at all times. This is possible only if a sufficiently long time is given at each decrement of temperature. Therefore equilibrium freezing requires infinitesimally slow rates of heat extraction and thus with ordinary cooling rates certain departures from equilibrium are to be expected.

Consider the cooling of the liquid solution of s in the following diagram.

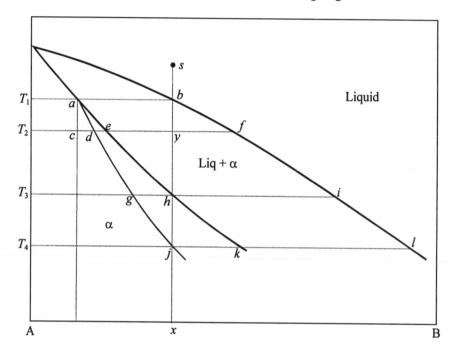

- The liquid remains homogeneous upon cooling from s to b.
- Freezing begins at T_1 with the deposition of crystals of the composition a.
- As cooling proceeds, the liquid composition changes along the liquidus.
- Solid forming at T_2 will have the composition of e and the liquid has the composition of f.
- During cooling from T_1 to T_2, a number of new nuclei form, each with the composition given by the solidus ae at its formation temperature. These nuclei will grow at the expense of the liquid.
- At the temperature T_2, none other than crystals having the composition e is in equilibrium with the liquid. However, not enough time is available for the compositions within the solid to change fully to the equilibrium values. Therefore the solid will have compositions ranging from c to e with an average somewhere between the two, say, d.
- On further cooling the liquid composition changes along the liquidus fi, the equilibrium solid composition along the solidus eh, and the average composition of the solid in a real practice along the line dg.

- Solidification would be completed at T_3, should equilibrium be maintained during the cooling. In reality, however, there is still liquid remaining.

$$\text{Fraction of liquid remained} = \frac{gh}{gi}$$

- On further cooling, the liquid composition changes along the liquidus il and the average solid composition along the line gj.
- At T_4, solidification is completed. It is thus seen that nonequilibrium freezing is characterised by
  1) increased temperature range over which liquid and solid are present
  2) $T_1 \leftrightarrow T_3$ for equilibrium solidification
  3) $T_1 \leftrightarrow T_4$ for nonequilibrium solidification
  4) a composition range remaining in the solids (at least $j \leftrightarrow k$).
- In summary, the solidification begins with the formation of numerous solid nuclei and the growth of these nuclei follows. Each nucleus has a gradient of composition from its centre to the periphery. This nonequilibrium effect is referred to as *coring*.

Example 2

Consider the A-B binary system. If A and B form a random solid solution with, say, 10 atom percent B, the probability of finding a B atom on any specific lattice site is just 0.1. Under certain conditions, however, B atoms may favour certain specific sites than the rest. B atoms will then preferentially position themselves on these specific sites. The probability of finding B atoms in these sites will greatly increase. This type of arrangement is referred to as an *ordered structure*. The process in which a random disordered solid solution is rearranged into an ordered solid solution is called an *order-disorder transition*.

Discuss the order-disorder transition using the phase diagram given :

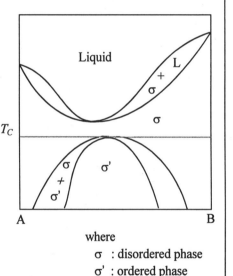

where
  σ : disordered phase
  σ' : ordered phase

- Ordered structure in the σ' field
- Disordered structure in the σ field
- Coexistence of ordered and disordered structures in the (σ + σ') field
- Disordered structure only at temperatures above T_C

Exercises

1. Consider the A-B binary peritectic system as shown in the following diagram :

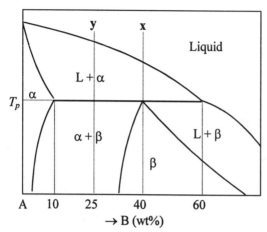

When liquid x is cooled maintaining equilibrium conditions, calculate wt% α that exists just above the peritectic temeperature T_p. When liquid y is cooled maintaining equilibrium conditions, calculate wt% β that exists just below the peritectic temperature T_p.

2. Phase diagram is one way of expressing thermodynamic equilibrium of a system. Every part of a phase diagram therefore has to conform to thermodynamic principles. Find errors in the phase diagrams shown below. Justify your answer.

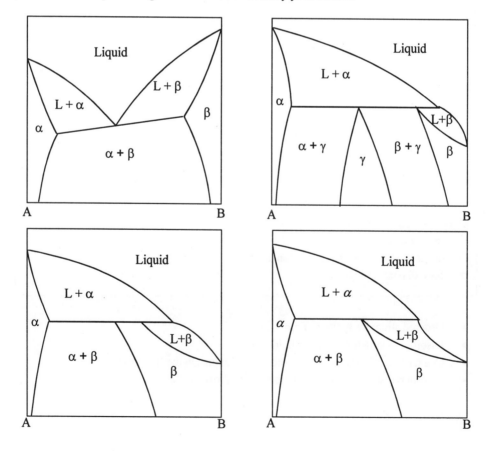

4.2.4. Thermodynamic Models

Ideal Solutions

Recall that the partial molar free energy, or chemical potential, of the component i in a solution is given by

$$\overline{G}_i = G_i^o + RT \ln a_i$$

If the solution behaves ideally over the entire range of composition,

$$a_i = N_i$$

$$\overline{G}_i = G_i^o + RT \ln N_i$$

Consider the A-B binary system consisting of liquid solution and α solid solution phases.

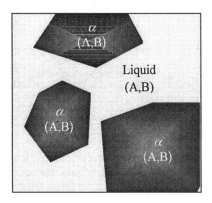

Chemical potentials of components of the liquid and phase are given as follows:

Liquid

$$\overline{G}_{A(L)} = G_{A(L)}^o + RT \ln N_{A(L)}$$
$$\overline{G}_{B(L)} = G_{B(L)}^o + RT \ln N_{B(L)}$$

α phase

$$\overline{G}_{A(\alpha)} = G_{A(S)}^o + RT \ln N_{A(\alpha)}$$
$$\overline{G}_{B(\alpha)} = G_{B(S)}^o + RT \ln N_{B(\alpha)}$$

Recall that at equilibrium the chemical potential of a component must be the same in all phases throughout the system. Therefore

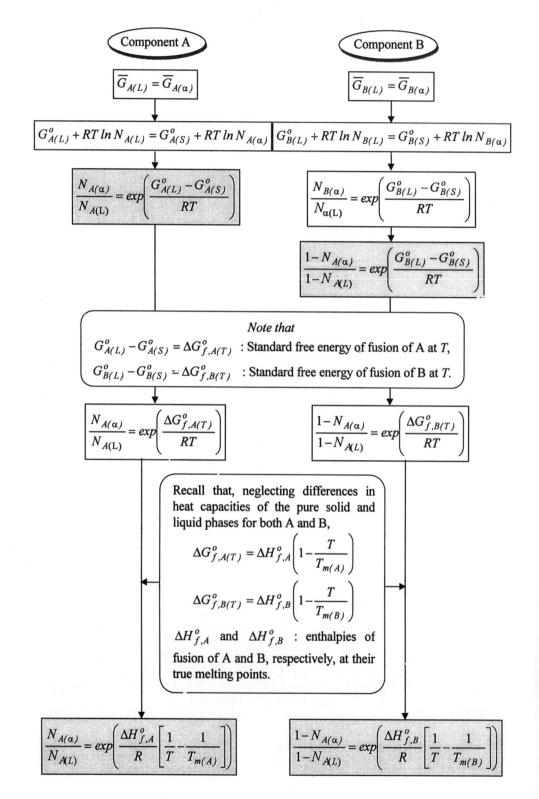

Thus in the ideal solution model of a two phase field, a knowledge of the enthalpies of fusion of the pure components at their respective melting points allows simultaneous solution of these two equations for the two unknowns, $N_{A(L)}$ and $N_{A(\alpha)}$ at the temperature of interest.

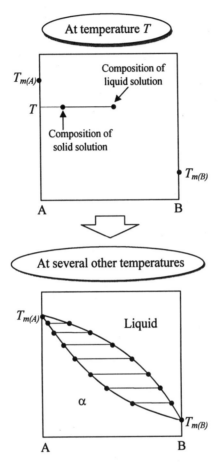

Given below is a typical form of phase diagram for an ideal binary solution:

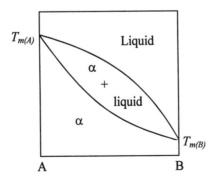

Pattern of a phase diagram depends on values of the enthalpies of fusion of the components in the solution, i.e., $\Delta H^o_{f,A}$ and $\Delta H^o_{f,B}$, and some examples are shown below :

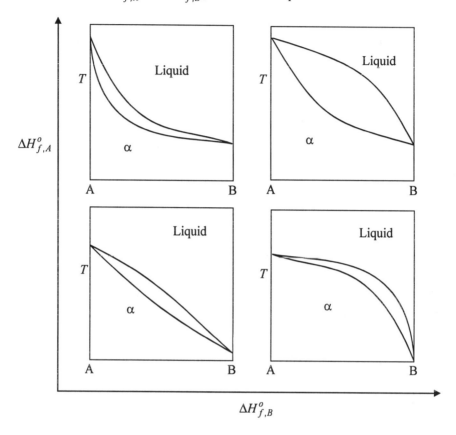

Non-ideal Solutions

Recall that the partial molar free energy, or chemical potential, of the component i in a solution is given by

$$\overline{G}_i = G^o_i + RT \ln a_i$$

Consider the A-B binary system consisting of α and β solid solutions.

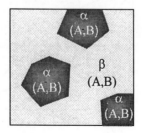

Chemical potentials of the components in α and β phases are

α phase

$$\overline{G}_{A(\alpha)} = G^o_{A(\alpha)} + RT \ln a_{A(\alpha)}$$
$$\overline{G}_{B(\alpha)} = G^o_{B(\alpha)} + RT \ln a_{B(\alpha)}$$

β phase

$$\overline{G}_{A(\beta)} = G^o_{A(\beta)} + RT \ln a_{A(\beta)}$$
$$\overline{G}_{B(\beta)} = G^o_{B(\beta)} + RT \ln a_{B(\beta)}$$

Recall that at equilibrium the chemical potential of a component must be the same in all phases throughout the system. Therefore,

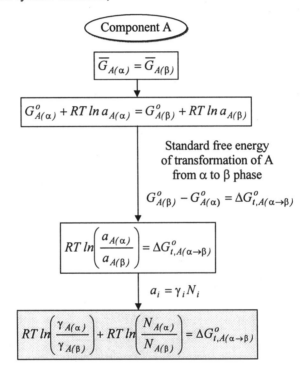

Component A

$$\overline{G}_{A(\alpha)} = \overline{G}_{A(\beta)}$$

$$G^o_{A(\alpha)} + RT \ln a_{A(\alpha)} = G^o_{A(\beta)} + RT \ln a_{A(\beta)}$$

Standard free energy of transformation of A from α to β phase

$$G^o_{A(\beta)} - G^o_{A(\alpha)} = \Delta G^o_{t,A(\alpha \to \beta)}$$

$$RT \ln\left(\frac{a_{A(\alpha)}}{a_{A(\beta)}}\right) = \Delta G^o_{t,A(\alpha \to \beta)}$$

$$a_i = \gamma_i N_i$$

$$RT \ln\left(\frac{\gamma_{A(\alpha)}}{\gamma_{A(\beta)}}\right) + RT \ln\left(\frac{N_{A(\alpha)}}{N_{A(\beta)}}\right) = \Delta G^o_{t,A(\alpha \to \beta)}$$

In a similar way for the component B,

$$RT \ln\left(\frac{\gamma_{B(\alpha)}}{\gamma_{B(\beta)}}\right) + RT \ln\left(\frac{1-N_{A(\alpha)}}{1-N_{A(\beta)}}\right) = \Delta G^o_{t,B(\alpha \to \beta)}$$

We now have two equations for two unknowns, $N_{A(\alpha)}$ and $N_{A(\beta)}$, provided that the standard free energies of transformations are known and the activity coefficients are given or expressed in terms of compositions. The values of $N_{A(\alpha)}$ and $N_{A(\beta)}$ thus found from the equations are the phase boundary compositions.

As real solutions may depart from ideality in a number of different ways, the activity coefficients in the above equations may take various expressions. We here discuss one simple type of departure from ideality, namely, the *regular solution model*, for the purpose of illustration.

For the A-B binary regular solution, we have seen in the section 2.3.5 that

$$RT \ln \gamma_A = \Omega N_B^2 \qquad RT \ln \gamma_B = \Omega N_A^2$$

Combination yields,

$$\boxed{\Omega_{(\alpha)}(1 - N_{A(\alpha)})^2 - \Omega_{(\beta)}(1 - N_{A(\beta)})^2 + RT \ln\left(\frac{N_{A(\alpha)}}{N_{A(\beta)}}\right) = \Delta G^o_{t,A(\alpha \to \beta)}}$$

$$\boxed{\Omega_{(\alpha)} N_{A(\alpha)}^2 - \Omega_{(\beta)} N_{A(\beta)}^2 + RT \ln\left(\frac{1 - N_{A(\alpha)}}{1 - N_{A(\beta)}}\right) = \Delta G^o_{t,B(\alpha \to \beta)}}$$

Parameters which are the input components for the calculation of phase boundary compositions, $N_{A(\alpha)}$ and $N_{A(\beta)}$, in the above equations are

$\Omega_{(\alpha)}$: Interaction parameter for the α phase standard state
$\Omega_{(\alpha)}$: Interaction parameter for the β phase standard state
$\Delta G^o_{t,A(\alpha \to \beta)}$: Standard free energy of phase transformation of A from α to β
$\Delta G^o_{t,B(\alpha \to \beta)}$: Standard free energy of phase transformation of B from α to β

The first two parameters ($\Omega_{(\alpha)}$ and $\Omega_{(\beta)}$) are independent of temperature, and the last two are constant at a given temperature. Thus all of these parameters are constant at any given temperature. The procedure of the phase boundary calculation will thus be represented by the following algorithm :

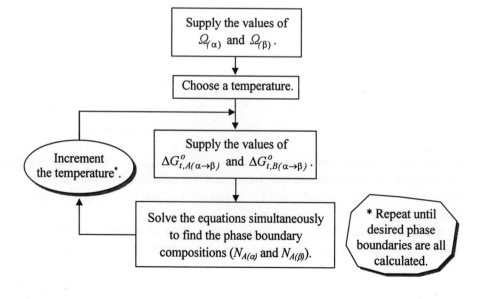

The regular solution model is much more flexible than the ideal solution model and a wide variety of phase diagrams can be produced with this model. Topographical changes in the phase diagram for a system A-B with regular solid and liquid phases brought about by systematic changes in the interaction parameters, $\Omega_{(s)}$ and $\Omega_{(l)}$, are illustrated in the following.* Melting points of pure A and B are assumed 800K and 1,200K, respectively, and standard entropies of fusion of A and B are assumed to be 10 J/mol K.

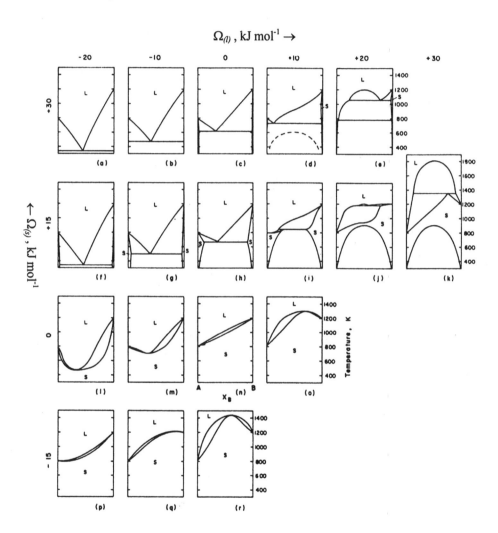

* Pelton, A.D. and Thompson, W.T.: Prog. Solid State Chem. 10, part 3, 1975, p119

Example 1

Prove that, if the heat capacity of pure liquid phase is the same as that of pure solid phase for both A and B in the A-B binary system, the equations of the liquidus and solidus are

$$\ln\left(\frac{\gamma_{A(\alpha)}}{\gamma_{A(L)}}\right) + \ln\left(\frac{N_{A(\alpha)}}{N_{A(L)}}\right) = \frac{\Delta H^o_{f,A(T_m)}}{R}\left(\frac{1}{T} - \frac{1}{T_{m(A)}}\right)$$

$$\ln\left(\frac{\gamma_{B(\alpha)}}{\gamma_{B(L)}}\right) + \ln\left(\frac{N_{B(\alpha)}}{N_{B(L)}}\right) = \frac{\Delta H^o_{f,B(T_m)}}{R}\left(\frac{1}{T} - \frac{1}{T_{m(B)}}\right)$$

$$RT\ln\left(\frac{\gamma_{A(\alpha)}}{\gamma_{A(L)}}\right) + RT\ln\left(\frac{N_{A(\alpha)}}{N_{A(L)}}\right) = \Delta G^o_{f,A(T)}$$

At T
$$\Delta G^o_{f,A(T)} = \Delta H^o_{f,A(T)} - T\Delta S^o_{f,A(T)}$$

At the melting point, $T_{m(A)}$
$$\Delta G^o_{f,A(T_m)} = 0 = \Delta H^o_{f,A(T_m)} - T_{m(A)}\Delta S^o_{f,A(T_m)}$$

$$\Delta S^o_{f,A(T_m)} = \frac{\Delta H^o_{f,A(T_m)}}{T_{m(A)}}$$

If $C_{P,A(L)} = C_{P,A(S)}$
$$\Delta H^o_{f,A(T)} = \Delta H^o_{f,A(T_m)}$$
$$\Delta S^o_{f,A(T)} = \Delta S^o_{f,A(T_m)}$$

$$\Delta G^o_{f,A(T)} = \Delta H^o_{f,A(T_m)}\left(1 - \frac{T}{T_{m(A)}}\right)$$

$$\ln\left(\frac{\gamma_{A(\alpha)}}{\gamma_{A(L)}}\right) + \ln\left(\frac{N_{A(\alpha)}}{N_{A(L)}}\right) = \frac{\Delta H^o_{f,A(T_m)}}{R}\left(\frac{1}{T} - \frac{1}{T_{m(A)}}\right)$$

For B, in a similar way,

$$\ln\left(\frac{\gamma_{B(\alpha)}}{\gamma_{B(L)}}\right) + \ln\left(\frac{N_{B(\alpha)}}{N_{B(L)}}\right) = \frac{\Delta H^o_{f,B(T_m)}}{R}\left(\frac{1}{T} - \frac{1}{T_{m(B)}}\right)$$

If the assumption that the heat capacities of the pure liquid and solid A are equal is unwarranted, the expression for the standard free energy of fusion of A at T must be corrected:

$$\Delta G^o_{f,A(T)} = \Delta H^o_{f,A(T)} - T\Delta S^o_{f,A(T)}$$

$$\Delta H^o_{f,A(T)} = H^o_{A(L),T} - H^o_{A(S),T}$$
$$\Delta S^o_{f,A(T)} = S^o_{A(L),T} - S^o_{A(S),T}$$

$$\Delta H^o_{f,A(T)} = \Delta H^o_{f,A(T_m)} + \int_{T_{m(A)}}^{T} \Delta C_{P,A} dT$$

$$\Delta S^o_{f,A(T)} = \Delta S^o_{f,A(T_m)} + \int_{T_{m(A)}}^{T} \frac{\Delta C_{P,A}}{T} dT$$

where

$$\Delta C_{P,A} = C_{P,A(L)} - C_{P,A(S)}$$

$$\Delta G^o_{f,A(T)} = \left(\Delta H^o_{f,A(T_m)} + \int_{T_{m(A)}}^{T} \Delta C_{P,A} dT \right) - T\left(\Delta S^o_{f,A(T_m)} + \int_{T_{m(A)}}^{T} \frac{\Delta C_{P,A}}{T} dT \right)$$

Exercises

1. Metals A and B behave ideally in both liquid and solid solutions. Calculate compositions of the solid and liquid solutions in equilibrium at 1,250K. The following information is known:

    A : T_m = 1,350K $\Delta H^o_{f,A} = 1,500 \text{Jmol}^{-1}$
    B : T_m = 700K $\Delta H^o_{f,B} = 3,300 \text{Jmol}^{-1}$

2. Components A and B behave regularly in both liquid and solid solutions. Find the compositions of the liquid and solid solutions in equilibrium at 1,300K. The following data are available:

    A : T_m = 1,350K $\Delta H^o_{f,A} = 9,330 \text{Jmol}^{-1}$
    B : T_m = 1,150K $\Delta H^o_{f,B} = 9,760 \text{Jmol}^{-1}$
    $\Omega_{(L)} = -5,600 \text{Jmol}^{-1}$ $\Omega_{(S)} = -11,400 \text{Jmol}^{-1}$

3. The A-B binary system behaves ideally in both its liquid and solid solutions. The element A melts at 1,500K with the heat of fusion of 14,700 Jmol^{-1}, and the heat

capacity difference, $\Delta C_{P,A}$ ($= C_{P,A(L)} - C_{P,A(S)}$), is approximately constant and equal to 4.6 J mol^{-1}K^{-1}. The element B melts at 2,300K, but the heat of fusion is unknown. The heat capacities of the pure liquid and solid B are equal. It was found in an experiment with a liquid solution of $N_B = 0.22$ that in cooling the first solid crystals appeared at 1,700K. Calculate the heat of fusion of B.

4. A and B have negligible mutual solid solubilities in the solid state and their phase diagram shows a eutectic transformation. The liquid phase at 1 atm is represented by the equation

$$G^E = 2,100 N_A N_B \left(1 - \frac{T}{3000}\right) Jmol^{-1}$$

The following information is known :

A: $T_m = 1,500$K $\Delta H^o_{f,A} = 15,100 J\,mol^{-1}$ $\Delta C_{P(S \to L)} = 0$

B: $T_m = 1,200$K $\Delta H^o_{f,B} = 11,100 J\,mol^{-1}$ $\Delta C_{P(S \to L)} = 0$

Develop equations of the liquidus of the above system and explain how to find the eutectic temperature and composition using the equations developed.

4.3. Ternary Systems

4.3.1. Composition Triangles

Ternary systems are those possessing three components. Therefore there are four independent variables in the A-B-C ternary system :
- Temperature
- Pressure
- Two composition variables (Third one is not independent since the sum of the mole (or mass) fractions is unity : $N_A + N_B + N_C = 1$).

Construction of a complete diagram which represents all these variables would require a four-dimensional space. However, if the pressure is assumed constant (customarily at 1 atm), the system can be represented by a three dimensional diagram with three independent variables, i.e., temperature and two composition variables. In plotting three dimensional diagrams, it is customary that the compositions are represented by triangular coordinates in a horizontal plane and the temperature in a vertical axis.

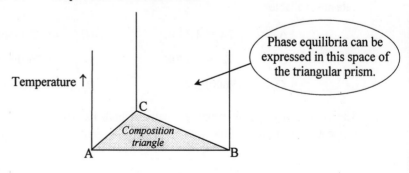

For plotting ternary compositions, it is common to employ an *equilateral composition triangle* with coordinates in terms of either mole fraction or weight percent of the three components.

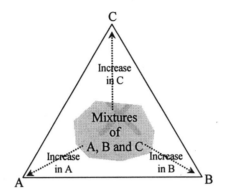

- Three pure components are represented by the apices, A, B and C.
- Binary compositions are represented along the edges : e.g., A point on the line B-C is composed entirely of components B and C without A.
- Points inside the triangle represent mixtures of all three components.

We now discuss several different methods of determining the proportions of three components represented by a point in the triangle.

> To determine the composition of the mixture represented by the point P in the following figures,

---

Method 1

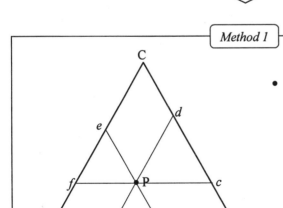

- Draw lines through P parallel to each of the sides of the triangle.

$$\text{Proportion of A} = \frac{Bb}{AB} = \frac{eC}{AC}$$

$$\text{Proportion of B} = \frac{Aa}{AB} = \frac{dC}{BC}$$

$$\text{Proportion of C} = \frac{Af}{AC} = \frac{cB}{BC}$$

Method 2

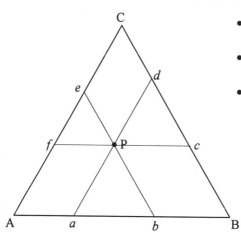

- Draw lines through P parallel to each of the sides of the triangle.
- Notice that each side is now divided into three parts.
- If the side A-B is chosen,

$$\text{Proportion of A} = \frac{bB}{AB}$$

$$\text{Proportion of B} = \frac{Aa}{AB}$$

$$\text{Proportion of C} = \frac{ab}{AB}$$

The composition can also be found from the sides B-C and A-C. The two end parts of each line represent the proportions of the components at the opposite ends and the middle part represents the proportion of the third component.

Method 3

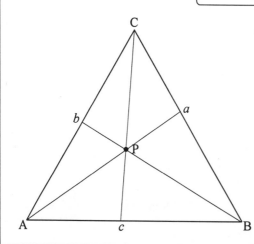

- Draw lines from apices through P to the opposite sides of the triangle.

$$\frac{\text{Proportion of A}}{\text{Proportion of B}} = \frac{cB}{Ac}$$

$$\frac{\text{Proportion of B}}{\text{Proportion of C}} = \frac{Ca}{aB}$$

$$\frac{\text{Proportion of C}}{\text{Proportion of A}} = \frac{Ab}{bC}$$

All the methods presented above are based on the same principle: i.e., the material balance using the *lever rule*. Therefore these are not limited to equilateral triangles, but equally valid for scalene triangles which frequently appear when dealing with subsystems.

Consider the subsystem XYZ within the system ABC.

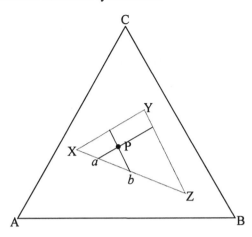

- Points X, Y and Z are mixtures of A, B and C and the composition of each point can be determined by one of the methods described above.
- Since the point P is inside the subsystem XYZ, it may be considered as a mixture of X, Y and Z :

$$\text{Proportion of X} = \frac{bZ}{XZ}$$

$$\text{Proportion of Y} = \frac{ab}{XZ}$$

$$\text{Proportion of Z} = \frac{Xa}{XZ}$$

Another important relationship which can be drawn from composition triangles is that

"If any two mixtures (or solutions) or components are mixed together, the composition of the resultant mixture lies on the straight line which joins the original two compositions."

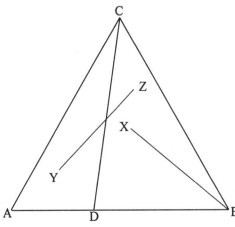

- If the component C is added to the binary mixture D, the composition of the resultant ternary mixture lies on the line CD.
- If the component B is added to the ternary mixture X, the composition of the mixture lies on the line XB.
- If the two ternary mixtures, Y and Z, are mixed together, the composition of the mixture lies on the line YZ.
- In all cases, the position of the resultant mixture on the line is determined by the lever rule.

As explained earlier, the temperature is represented by an axis perpendicular to the plane of the composition triangle. The point S in the diagram below left represents a ternary mixture of the composition P at the temperature T_1. (Recall that the pressure is assumed constant in this type of composition-temperature coordinates.). The diagram below right is an example of the three dimensional ternary phase diagram drawn using the composition triangle - temperature coordinates.

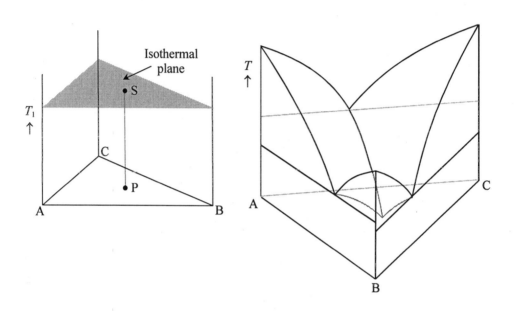

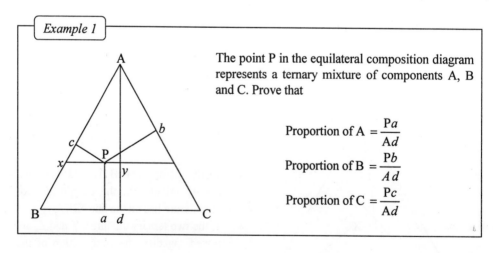

Proportion of A $= \dfrac{Bx}{AB} = \dfrac{yd}{Ad} = \dfrac{Pa}{Ad}$. The proportions of B and C can be obtained in a similar way.

Example 2

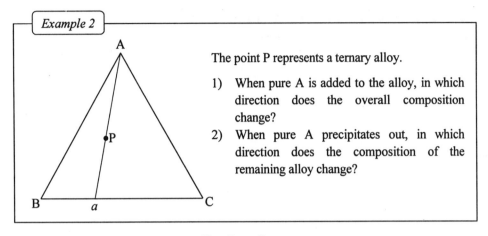

The point P represents a ternary alloy.

1) When pure A is added to the alloy, in which direction does the overall composition change?
2) When pure A precipitates out, in which direction does the composition of the remaining alloy change?

1) From P to A
2) From P to a

Example 3

The Gibbs phase rule is of use in phase equilibrium studies of multicomponent systems.

1) Determine the maximum number of phases which can coexist in equilibrium in a ternary system.
2) For condensed systems the effect of pressure is negligible in many cases. Therefore, when the pressure is fixed at 1 atm, the number of variables of the system is reduced by one unit. When equilibrium conditions of a ternary system at a constant pressure are represented in the space of the composition triangle - temperature prism, prove the following:
   a) Four phase equilibria are represented by points.
   b) Three phase equilibria are represented by lines.
   c) Two phase equilibria are represented by surfaces.
   d) Single phase equilibria are represented by spaces.

1) From the Gibbs phase rule,

$$f = c - p + 2$$

$c = 3$ (ternary system)
$f = 0$ (the maximum number of phases occurs at zero degree of freedom.)

$$p = 5$$

A maximum of five phases can coexist in equilibrium.

2) If the pressure is fixed at a constant value,

$$f = c - p + 1$$

	p	c	f	Explanation
a)	4	3	0	No degree of freedom : Neither composition nor temperature can be chosen freely. In other words four phases can exist together in equilibrium only at a fixed composition and temperature : *invariant*
b)	3	3	1	One degree of freedom : One variable (either the concentration of one of the components or temperature) can be freely varied, and then all others are fixed : *univariant*.
c)	2	3	2	Two degrees of freedom : Two variables are at our discretion, and the rest are then fixed : *bivariant*.
d)	1	3	3	Three degrees of freedom : Three variables can be varied freely: e.g., After choosing a composition of the ternary system, temperature can still be varied while maintaining single phase state : *trivariant*.

Exercises

1. Fig. A below shows the composition triangle of the ABC ternary system. Determine the composition of the mixture represented by the point P.

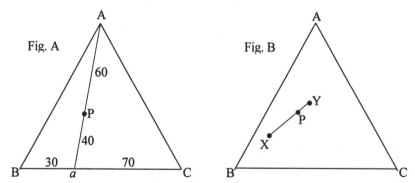

2. A mixture represented by the point P in Fig. B above is to be prepared by mixing the mixtures X and Y. Determine the ratio of X to Y to obtain the right composition. The composition of each point is given in the following table:

	A	B	C
P	35%	40%	25%
X	20%	70%	10%
Y	40%	30%	30%

4.3.2. Polythermal Projections

This figure represents a simple ternary phase diagram. However, it has the disadvantage that lines in the figure are not seen in true length, and hence it is difficult to obtain quantitative information.

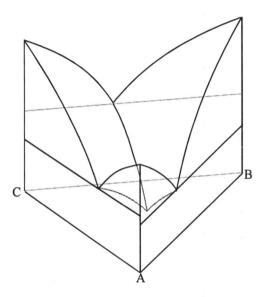

There are two ways to solve this problem:
1) A two dimensional representation of the ternary liquidus surface on the base composition triangle.
2) Two dimensional isothermal diagrams which represent isothermal plane intersections with various surfaces (liquidus, solidus, etc).

The first method consists of a *polythermal projection* of all features (liquidus, etc) down onto the base composition triangle.

The figure below shows such a polythermal projection of a simple ternary eutectic system without solid solution like the one represented by the figure above.

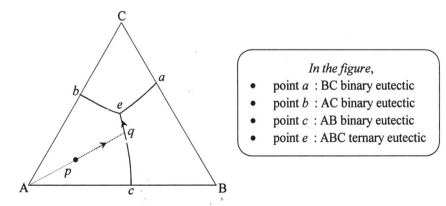

In the figure,
- point a : BC binary eutectic
- point b : AC binary eutectic
- point c : AB binary eutectic
- point e : ABC ternary eutectic

Now, let's examine crystallisation paths of this simple ternary system. If a liquid of composition *p* is allowed to cool,

- The liquid solution remains liquid until the system temperature reaches the liquidus.
- At the liquidus, pure solid A begins to crystallise.
- As the temperature decreases further, solid A continues to precipitate out of the liquid, and hence the liquid is depleted in A and the liquid composition changes along the line *pq*.
- At *q*, the second phase B appears, and the liquid composition moves along the curve *qe*. Until it reaches the ternary eutectic point *e*, both A and B crystallise.
- At *e*, solid phases A, B and C crystallise and the temperature remains constant until all liquid has exhausted.
- The final product will consist of large crystals of A and B which have crystallised before reaching the point *e*, and small crystals of eutectic structure of A, B and C which have crystallised at the point *e*.

The polythermal projection is generally given with constant temperature lines as shown in the following figure:

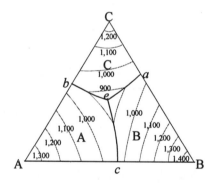

These lines are called *liquidus isotherms*. The intersections of adjoining liquidus surfaces like *ae*, *be* and *ce* are called the *boundary curves*. When a liquid whose composition lies in the region surrounded by A*ceb* is cooled, the first crystalline phase that appears is A, and hence A is called the *primary phase* and the region A*ceb* is the *primary field* of A. In this field, solid A is the last solid to disappear when any composition within this field is heated. Similarly, B and C are primary phases in their respective primary fields, B*aec* and C*aeb*.

In multicomponent systems, compounds are frequently formed between components. The following phase diagrams are for ABC ternary system forming a binary compound AB which melts *congruently*, as it is stable at its melting point :

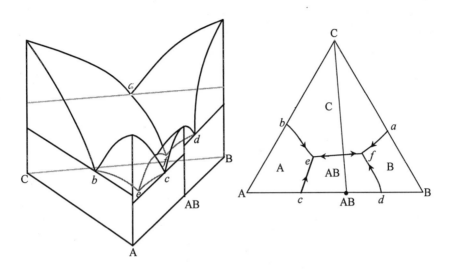

The straight line C-AB is called an *Alkemade line*. An Alkemade line divides a ternary composition triangle into two sub-composition triangles. The final phases produced by equilibrium crystallisation of any composition within one of these sub-triangles are those indicated by the apices of the triangle. For instance, any composition within the composition triangle A-C-AB results in producing phases A, C and AB at equilibrium. The crossing point on the boundary curve *ef* by the Alkemade line is the maximum in temperature on the curve *ef*, and the points *e* and *f* are eutectic, and each sub-composition triangle can be treated as a true ternary system. The Alkemade line in this case represents a true binary system of C and AB. The arrows in the diagram indicate directions of decreasing temperature.

The following phase diagrams are for ABC ternary system forming a binary compound AB which melts *incongruently*, as it is unstable at its melting point :

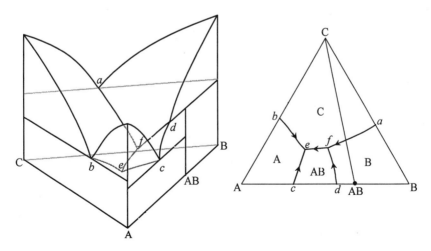

In this case, the Alkemade line C-AB does not cross the boundary curve (*ef*) between these primary phases. Now, we state the *Alkemade theorem* in a more general form :

The direction of falling temperature on the boundary curve of two intersecting primary phase areas is always away from the Alkemade line.
- If the Alkemade line intersects the boundary curve, the point of intersection represents a temperature maximum on the boundary curve.
- If the Alkemade line does not intersect the boundary curve, then temperature maximum on the curve is represented by that end which if prolonged would intersect the Alkemade line.

Now, it is obvious from the Alkemade theorem that for the case of the incongruently-melting compound as shown above only the point *e* is eutectic. It is also apparent by examining the above phase diagrams for the congruently-melting compound and the incongruently-melting compound that the composition of the compound lies
- *within the primary field* of the compound if it has a congruent melting point, and
- *outside the primary field* of the compound if it has an incongruent melting point.

The ternary invariant points (e.g., *e* and *f* in the above diagrams) that appear in a system without solid solution are either *ternary eutectics* or *ternary peritectics*. Whether it is eutectic or peritectic is determined by the directions of falling temperatures along the boundary curves.
- If an invariant point is the minimum point in temperature along all three boundary curves, it is a ternary eutectic.
- If the point is not the minimum point, it is a ternary peritectic.

In the previous diagrams, points *e* and *f* for the congruently-melting compound are both ternary eutectic. On the other hand, for the incongruently-melting compound the point *e* is the ternary eutectic, whereas the point *f* is the ternary peritectic.

Alkemade lines are also called in many different ways including *conjugation lines* and *joins*. Both Alkemade lines and *Alkemade triangles* (composition triangles produced by Alkemade lines) are of use in the understanding of ternary systems. They play an essential role in understanding crystallisation or heating paths :

Example 1

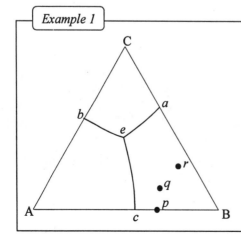

This figure is a simple ternary phase diagram which shows a ternary eutectic at the point *e*.

Explain the crystallisation path of each of liquids represented by the points *p*, *q*, and *r*. Explain also the change in (1) liquid composition, (2) mean solid composition, (3) solid phase(s), (4) instantaneous composition of solids crystallising, and (5) change in the ratio of liquid to solid phases.

As the whole diagram is an Alkemade triangle and point e is the ternary eutectic, all crystallisation curves of this ternary system should terminate at this ternary eutectic. However, binary liquid compositions such as points *p* and *s* terminate their respective binary eutectic, *c* and *a*.

1) *Point p*
   - The system remains liquid until the temperature reaches liquidus.
   - At the liquidus, solid B begins to crystallise.
   - As the temperature decreases, solid B continues to crystallise and the liquid composition changes toward point c along the line *pc*.
   - The ratio of solid (B) to liquid at the moment the liquid composition arrives at point *c* is represented by the lever rule, *p-B/c-p*.
   - At point *c*, both A and B co-crystallise forming eutectic structure until all liquid is consumed. The temperature remains constant during this eutectic reaction :

$$L = A_{(s)} + B_{(s)}$$

2) *Point q*
   - At the liquidus solid B begins to crystallise.
   - As the temperature decreases, solid B continues to crystallise and the liquid composition moves straight away from B along the straight line *qu*.
   - At point *u*, the second phase A appears.
   - With further cooling, the crystallisation path follows the boundary curve *ue* with crystallisation of both A and B.

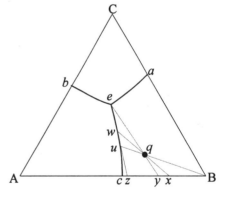

   - At point *w* which is an arbitrary point on the boundary curve *ue*, the mean composition of the solid is represented by the point *x*. Proportions of solid A and solid B are determined by the lever rule: (solid A)/(solid B) = *x-B/A-x*. The ratio of liquid to solid is also given by the lever rule - *q-x/w-q*.
   - The instantaneous composition of the solid phases crystallising at *w* can be determined by drawing the tangent to the curve *ue* at *w* and finding the intersection of the tangent with line AB. Therefore, point *z* represents the instantaneous composition of the solid phases.
   - At point e, the eutectic crystallisation occurs :

$$L = A_{(s)} + B_{(s)} + C_{(s)}$$

   - During the eutectic reaction, the mean composition of the solid phases changes along the line *yq*. The solid composition reaches point *q* when the liquid is completely consumed by the eutectic crystallisation.

- In summary, the liquid composition changes along the path *quwe*, whereas the mean composition of solid phases follows the path B*xyq*.
- The final structure of the system after complete solidification consists of large crystals of A and B which have been crystallised during the path *quwe* and a mixture of small crystals of A, B and C (eutectic structure).

3) *Point r*
   - The crystallisation path can be determined in the same manner as the above case. However, the second phase crystallising in this case is C instead of B.

Example 2

This figure is a phase diagram of ABC ternary system which forms two binary compounds, AC and BC.

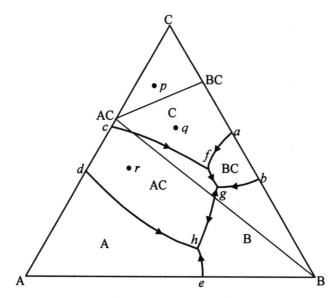

1) Why is an Alkemade line connecting A and BC not drawn?
2) Describe crystallisation paths of liquid compositions *p,q* and *r*.
3) Describe heating paths of mean solid compositions of *p,q* and *r*.
4) Would the compounds AC and BC melt congruently or incongruently?

---

1) Because the primary fields of A and BC are not in contact with each other, and hence these two fields do not form a boundary curve.

2) *Point p*

   As point *p* lies in the Alkemade triangle C-AC-BC, the final solid phases in equilibrium should be C, AC and BC. From the figure given below,

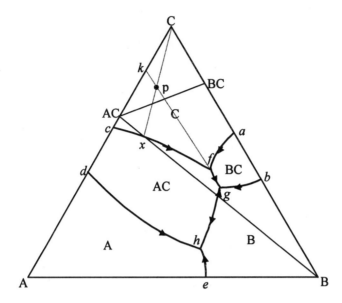

- On touching the liquidus at p, solid C begins to crystallise.
- From p to x, solid C continues to crystallise.
- At x, solid compound AC appears.
- From x to f, both AC and C crystallise.
- As the mean composition of the solid crystallised out between p and f is given by point k, there is some liquid left when crystallisation path arrives at f.
- At f, the final solidification takes place through a peritectic reaction : AC and BC crystallise out together at the expense of liquid and portion of C. This can easily be seen from the Alkemade triangle C-AC-BC. During the peritectic reaction, the mean solid composition moves from k to p.
- At completion of the solidification at point f, the final solid phases in equilibrium are C, AC and BC, since the mean composition of the system lies within the Alkemade triangle C-AC-BC as mentioned earlier.

Point q

As point q lies in the Alkemade triangle AC-BC-B, the final solid phases in equilibrium should be AC, BC and B. Refer to the figure given below:
- On touching the liquidus at q, solid C begins to crystallise.
- From q to x, solid C continues to crystallise.
- At x, solid AC appears.
- From x to f, AC and C crystallise out together.
- As the mean composition of the solid crystallised out between q and f is given by point k, there is some liquid left when crystallisation path arrives at f.
- At f, a peritectic reaction takes place: AC and BC crystallise out together at the expense of C and portion of liquid. During the peritectic reaction, the mean solid composition moves from k to m.

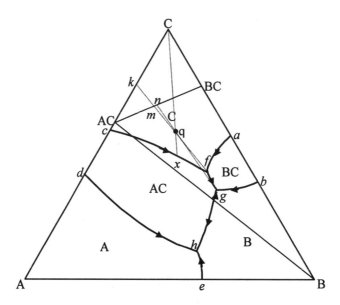

- From f to g, AC and BC crystallise out together, and the mean solid composition moves from m to n.
- At g, a eutectic reaction takes place until the last portion of liquid is completely consumed : three solid phases AC, BC and B crystallise out together to form a eutectic structure.
- During the eutectic reaction at g, the mean solid composition changes from n to q.
- At completion of the solidification at point g, the final solid phases in equilibrium are AC, BC and B since the mean composition of the system lies within the Alkemade triangle AC-BC-B as mentioned earlier.

Point r

As point r lies in the Alkemade triangle AC-A-B, the final solid phases in equilibrium should be AC, A and B.

Referring to the figure given below:
- On touching the liquidus at r, solid AC begins to crystallise.
- From r to z, solid AC continues to crystallise.
- At z, solid A appears.
- From z to h, both AC and A crystallise.
- As the mean composition of the solid crystallised out between r and h is given by point k, there is some liquid left when crystallisation path arrives at h.
- At h, the final solidification takes place through a eutectic reaction: AC, A and B crystallise out together to form a eutectic structure. During the eutectic reaction, the mean solid composition moves from k to r.

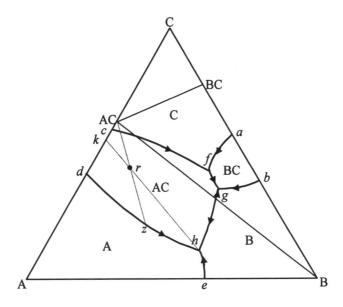

3) Heating from below the liquidus can be considered as the opposite to cooling from above the liquidus. Heating path of composition p is illustrated below. Others may be determined in the same manner.

Mean solid composition of p

- At the peritectic point f, the peritectic reaction of AC and BC reacting form both C and liquid phase proceeds until BC is consumed completely. This occurs isothermally. The composition of the liquid phase formed by this peritectic reaction is that of point f.

- Above the peritectic temperature, solids C and AC continue to dissolve into the liquid phase, and the liquid composition changes from f to x.

- When the liquid reaches point x, solid AC has completely dissolved and C is the only solid phase left.

- As the temperature increases, solid C continues to dissolve and the liquid composition moves from x to p.

- When the liquid reaches point p, solid C has completely dissolved and the system becomes a liquid phase of the composition of p.

4) Both AC and BC melt incongruently, as the stoichiometric compositions of these compounds lie outside their respective primary fields.

Example 3

Referring to the figure given below,
1) Determine whether binary compound δ and ternary compounds ε and φ melt congruently or incongruently.
2) Draw all Alkemade lines.
3) Determine the crystallisation path of a liquid of composition p.

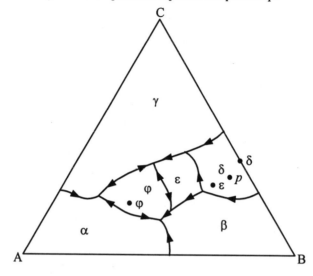

1) The binary compound δ and the ternary compound φ melt congruently, since they are within their respective primary fields, but the ternary compound ε melts incongruently as it is outside its primary field.

2) All the Alkemade lines are given in the following diagram.

3) Crystallisation path (Refer to the figure given below):

- When the temperature reaches the liquidus, δ phase begin to crystallise.
- Phase δ continues to crystallise out until the liquid composition contacts the boundary curve between δ and β at point k.
- At k, phase β appears, and thereafter δ and β crystallise together with the liquid composition changing along the path km.
- At point m, a ternary peritectic reaction occurs isothermally.

$$L + \delta + \beta \rightarrow \varepsilon$$

- There are three possible results after completion of the above peritectic reaction:
  1) The liquid is exhausted before either δ or β, and hence solidification is completed at point m.
  2) δ phase is exhausted first, and crystallisation continues along the path mo.
  3) β phase is exhausted first, and crystallisation continues along the path mn.

Which is the case depends on within which Alkemade triangle the total composition lies. Since point p lies within the triangle δ-ε-B(β), solidification should terminate at point m.

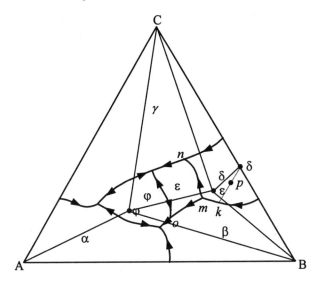

Exercises

1. Discuss the crystallisation paths of the overall liquid compositions p and q in the following ternary phase diagram:

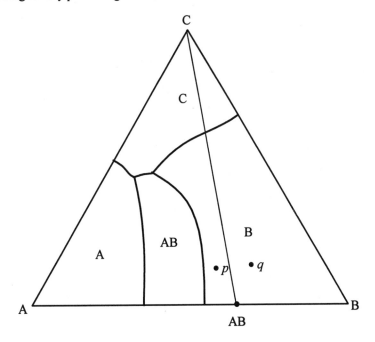

2. Shown below is the phase diagram of the SiO_2-CaO-Al_2O_3 system.* Discuss solidification paths for the compositions p, q and r indicated on the diagram.

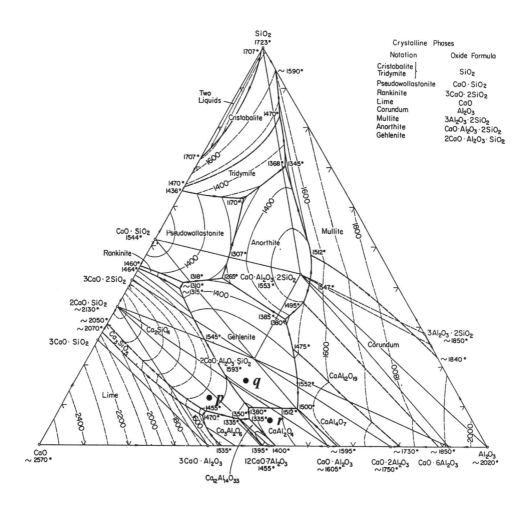

* *"Phase diagrams for ceramists"*, E.M. Levin, C.R. Robbins and H.F. McMurdie, The American Ceramic Society, Inc. (1964), p219

4.3.3. Isothermal Sections

The following figure represents a simple ternary phase diagram. However, it has the disadvantage that lines in the figure are not seen in true length, and hence it is difficult to obtain quantitative information.

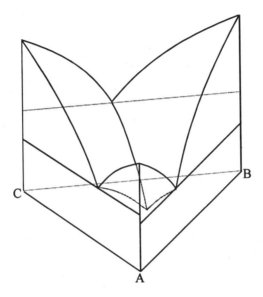

In order to solve this problem, the polythermal projection method was discussed in the previous section (Section 4.3.2). In this section, the other method, namely, two dimensional isothermal diagrams are discussed.

First, we discuss a simple ternary eutectic system without solid solution as shown above. The following figures show isothermal sections at a number of different temperatures :

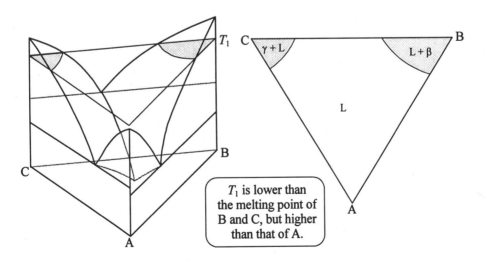

T_1 is lower than the melting point of B and C, but higher than that of A.

224   *Chemical Thermodynamics for Metals and Materials*

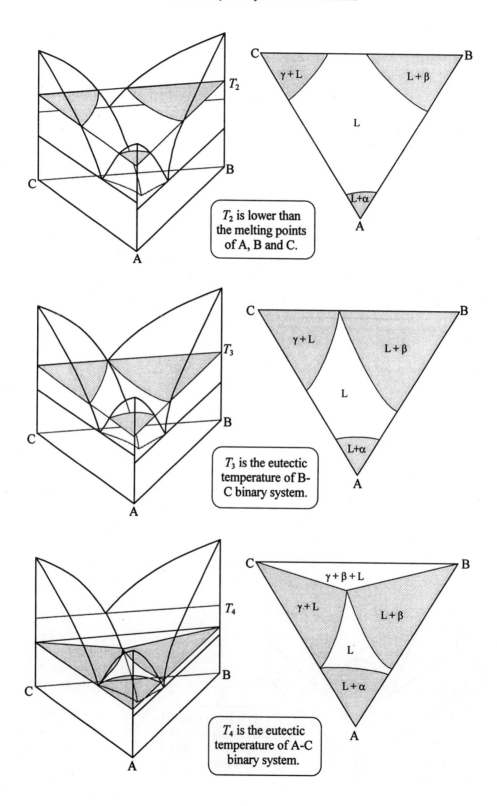

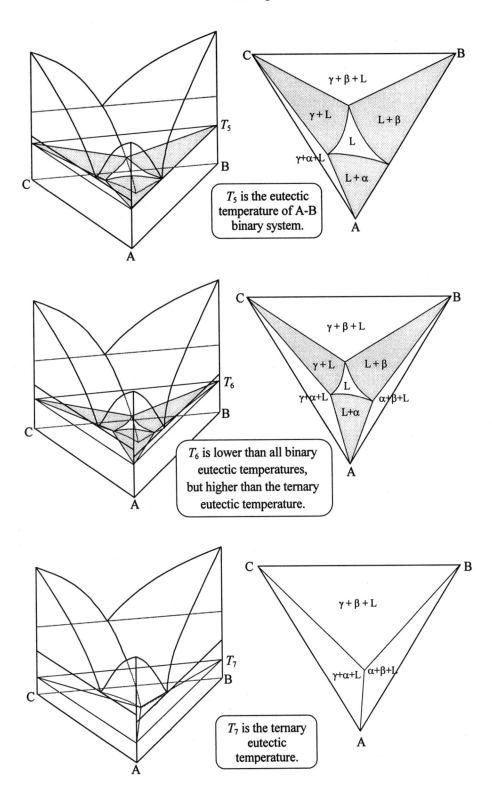

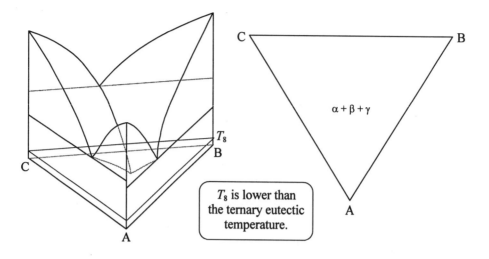

T_8 is lower than the ternary eutectic temperature.

Polythermal projections of the liquidus discussed in section 4.3.2. do not provide information on the compositions of solid phases if solid solutions or non-stoichiometric compounds are formed at equilibrium. For providing this information, the method of isothermal section is particularly useful. The following figure represents a simple ternary eutectic system with *terminal solid solutions* formed.

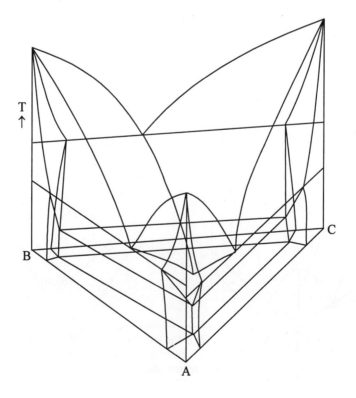

Shown in the following are the isothermal sections at a number of different temperatures of an hypothetical system ABC like the one given above :

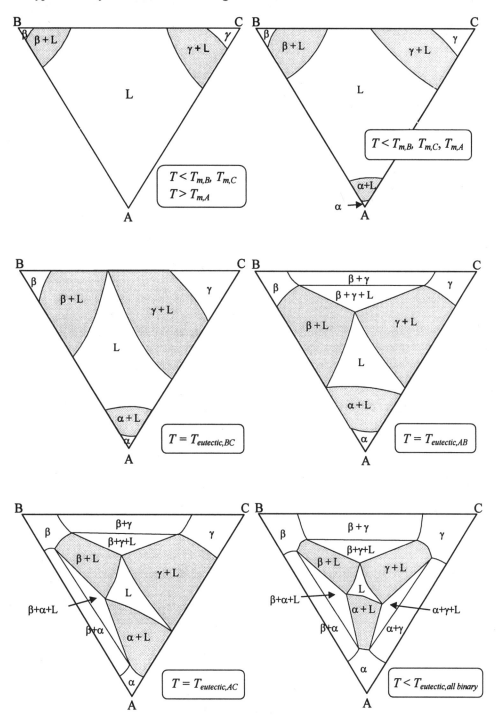

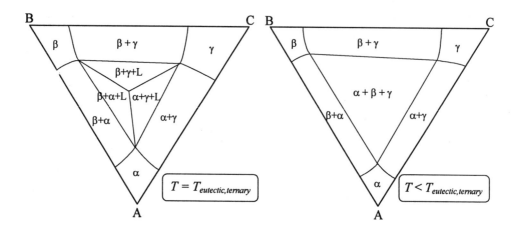

- If the overall composition of an alloy lies within the liquid phase region at a particular temperature, the alloy will exist as liquid.
- If the overall composition lies within one of the solid solution regions, the alloy will exist as the corresponding solid solution.
- If the composition lies within one of two-phase regions, say, β and liquid phases, both β and liquid phases will coexist.

Now a question arises as to how to determine the compositions of β phase and the liquid phase which are in equilibrium with each other. The usual practice is to include *tie lines* in the isothermal sections, which join the composition points of *conjugate phases* which coexist in equilibrium at a given temperature and pressure.

Tie lines are in fact common tangents to the Gibbs free energy surfaces of the phases that coexist in equilibrium. Let's assume that α and β phases, both of which form solid solutions, coexist. The following figure shows two arbitrary surfaces, α and β :

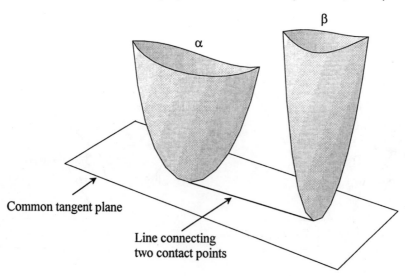

A common tangent plane shown above produces a pair of contact points. By rolling the plane on the surfaces, an infinite number of pairs of contact points are generated. Mapping on a reference plane of lines which connect selected conjugate pairs will show how contact points on one surface correspond to those on the other surface.

At a given temperature, the Gibbs free energy of each phase in a ternary system may be represented in a graphical form with the composition triangle as base and the free energy as vertical axis. Then it would look like this :

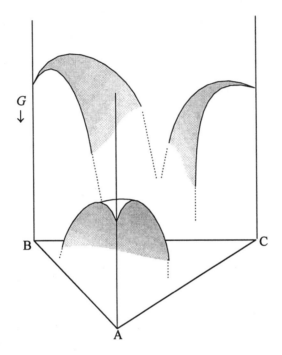

The phases that would exist at equilibrium and the compositions of the phases are determined by the contact points of a common tangent plane to their Gibbs free energy surfaces. Connection of the contact points forms a *tie line*. In a ternary system, a common tangent plane can contact two Gibbs free energy surfaces at an infinite number of points, and hence an infinite number of tie lines are generated. However, a common tangent plane to three Gibbs free energy surfaces generates only one set of contact points. Connection of these points forms a *tie-triangle*.

The following is an example of an isothermal section showing a number of tie lines :

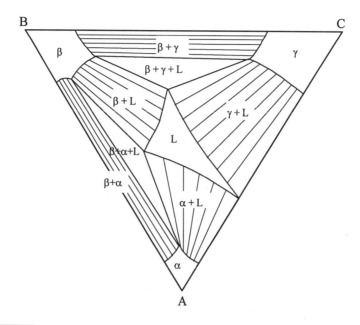

Example 1

In construction of isothermal sections or isothermal ternary phase diagrams a tie triangle together with contacting single phase and two-phase areas play an important role. In fact it may be considered as a building block of isothermal phase diagrams. Isothermal ternary phase diagrams are composed of a number of these building blocks. Shown below is an example of a tie-triangle with adjacent single and two phase areas :

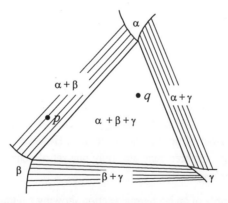

1. Determine relative proportion of each phase that exist at equilibrium for a system of overall composition of p in the diagram.
2. Determine the fraction of α phase that exist at equilibrium for the overall composition of q.
3. Are the boundary curves at the α phase corner correct?

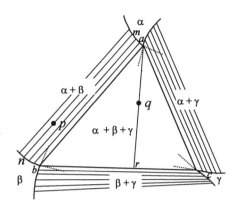

1) Fraction of α phase = $\dfrac{pn}{mn}$

2) Fraction of α phase = $\dfrac{qr}{ar}$

3) Both extensions of the boundary curves of the single phase areas must project either into the triangle (γ corner), or outside triangle (β corner), but not in mix. The α corner is wrong.

Exercises

1. Prove that tie lines must not cross each other within any two phase region.
2. ABC ternary system forms three binary eutectics and a ternary eutectic as shown below. Discuss equilibrium cooling paths for the overall compositions p, q and r indicated in the diagram. Discuss also the change in microstructure that should occur during cooling.

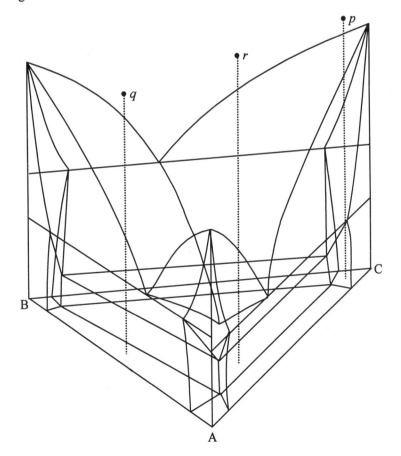

CHAPTER 5

ELECTROCHEMISTRY

5.1. Electrochemical Concepts and Thermodynamics

5.1.1. Basic Electrochemical Concepts

Reactions which involve electron transfer are of the general category of oxidation-reduction or *redox* reactions.

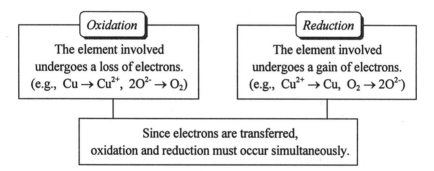

The electrons lost by one species are taken up by the other.

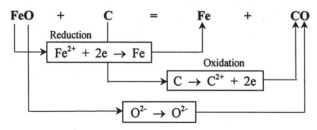

The stoichiometry of a reaction which involves electron transfer is related to the electrical quantities determined by *Faraday's law* which states
- One equivalent of product is produced by the passage of 96,487 coulombs of charge, or
- One equivalent of chemical change produces 96,487 coulombs of electricity.

One equivalent of a substance undergoing oxidation produces *Avogadro's number* (N), or one mole of electrons :

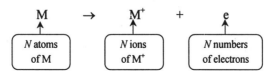

According to Faraday's law, therefore, the passage of 96,487 coulombs of electricity corresponds to the passage of Avogadro's number, $N = 6.023 \times 10^{23}$, of electrons. Faraday's law is understandable because an Avogadro's number of electrons added to or removed from a reagent will produce an equivalent of product. The quantity of charge that corresponds to a chemical equivalent is known as the *Faraday*, F:

$$F = 96,487 \text{ Coulombs}$$

Suppose a redox reaction takes place in such a way that the reaction produces a detectable electric current. This will be possible only if the reaction can be separated physically so that electrons are lost in one part of the system and gained in another.

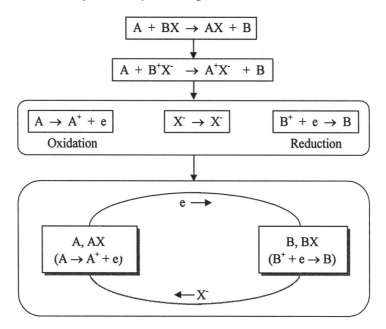

In the above device, a mixture of A and AX, and a mixture of B and BX are physically separated, but joined by two connections, i.e.,
- Electronic conductor which allows the passage of electrons only
- Ionic conductor which allows the passage of X^- ions only.

With the arrangement shown above, the reaction proceeds spontaneously, in which electrons move from left to right and X^- ions from right to left so that the electroneutrality is maintained. This type of reactions which take place in an electrochemical manner is called *electrochemical reaction*. A device like the one shown above, which permits a spontaneous electrochemical reaction to produce a detectable electric current, is termed a *galvanic cell*.

As shown in the above figure, oxidation occurs in one *half-cell* and reduction occurs in the other half-cell. The electrode at which oxidation occurs is referred to as the *anode*, while the electrode at which reduction occurs is termed *cathode*.

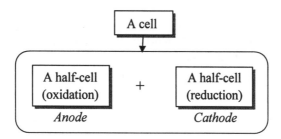

If the free energy change for the reaction

$$A + BX \rightarrow AX + B \qquad \Delta G_r$$

is negative, the reaction proceeds from left to right as written. As the electrochemical reaction is basically the same as the above chemical reaction (but in different arrangement in which electrons move from the anode to the cathode), the *driving forces* for the chemical reaction (ΔG_r) and for the electrochemical reaction (i.e., driving force for the electron movement) must be related to each other.

Electrons flow if there is an electric potential difference or an electrical voltage. If a galvanic cell is set up in conjunction with an appropriate external circuit so that the electric potential difference which exists in the cell is opposed by an identical, but opposite in direction, voltage from an external source, the electrochemical reaction of the cell ceases to proceed. The electrochemical reaction is now brought into equilibrium. This type of equilibrium is referred to as *electrochemical equilibrium*. That is, when the chemical driving force (ΔG_r) is balanced with the opposing electrical driving force, the reaction is in equilibrium electrochemically even though $\Delta G_r \neq 0$.

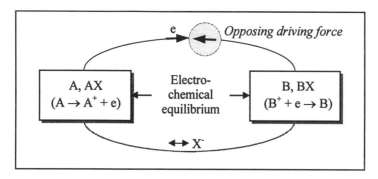

The application of thermodynamic principles to this type of electrochemical processes is discussed in the next section (5.1.2).

Exercises

1. Calculate the number of electrons released when one gram mole of calcium is oxidised. Calculate the coulombs of electricity generated by the oxidation.

2. The galvanic cell performs an electric work when the electrochemical reaction of the cell occurs. As work is not a state function, the amount of work which a system does depends on the path it takes for given initial and final states. Under what conditions does the amount of work the cell performs become maximum?

5.1.2. Electrochemical Cell Thermodynamics

Consider the reaction

$$A + BX \rightarrow AX + B \qquad \Delta G_r < 0$$

If the mixture of all reactants and products is in a same reactor, the above chemical reaction will proceed from left to right with a driving force equivalent to the free energy change of ΔG_r. The reaction will continue until $\Delta G_r = 0$. If a galvanic cell is constructed as shown below, the same reaction will occur, but in an electrochemical manner.

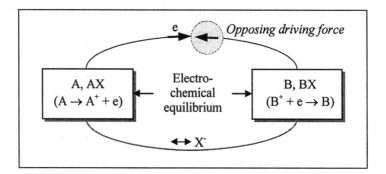

In the absence of the opposing driving force, the electron transfer from left to right, which enables the electrochemical reaction to occur, will enjoy the full driving force equivalent to ΔG_r. If an external force opposing to this driving force is applied, the net driving force for the electron transfer will be reduced accordingly. The electrochemical reaction of the cell will then suffer a decrease in the driving force. When the opposing external force (voltage) is increased to exactly balance the driving force of the electrochemical reaction of the cell, the electrons will cease to flow and so will the cell reaction.

Now we examine the amount of work the cell performs under various conditions. To help understand the concept of work of a cell, the case of gas expansion in a cylinder is reviewed below :

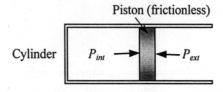

- If $P_{ext} = 0$, the piston moves without doing work because there is nothing to work against.
- If P_{ext} is increased, the piston moves with doing work against P_{ext}.
- If P_{ext} is only infinitesimally smaller than P_{int}, i.e., $P_{ext} = P_{int} - dP$, the piston moves with doing the maximum work, w_{max}, against P_{ext}.
- If P_{ext} is infinitesimally larger than P_{int}, i.e., $P_{ext} = P_{int} + dP$, the direction of the piston movement will be reversed.
- The last two cases depict the conditions for reversibility of gas expansion.

Work under reversible conditions (w_{rev}) = the maximum work (w_{max})

$$w_{rev} = w_{max}$$

Work of electrochemical cells has analogy with work of gas expansion discussed above.

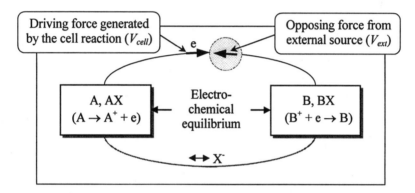

- If there is no opposing force from an external source, i.e., $V_{ext} = 0$, electrons will move without doing work because there is nothing to work against.

- If the external opposing force (V_{ext}) is increased, the electrons move with doing work against V_{ext}. The net driving force for the electron transfer (ΔV) is

$$\Delta V = V_{cell} - V_{ext}$$

- If V_{ext} is only infinitesimally smaller than V_{cell}, i.e.,

$$V_{ext} = V_{cell} - dV$$

the electrons move with doing maximum work, w_{max}, against V_{ext}. Since the net driving force for the electron transfer is infinitely small (dV), the cell reaction will proceed at an infinitely slow rate, but performing the maximum work.

- If V_{ext} is only infinitesimally larger than V_{cell}, i.e.,

$$V_{ext} = V_{cell} + dV$$

the direction of the electron transfer is reversed. In other words, the direction of the cell reaction can be reversed by an infinitesimal change in the external opposing force.

- The last two cases depict the conditions of reversibility of the electrochemical reaction of the cell.

Work under reversible conditions (w_{rev}) = the maximum work (w_{max})

$$w_{rev} = w_{max}$$

In general the *electromotive force* of a cell operating under reversible conditions is referred to as *emf* ($\mathcal{E}$), while that observed when conditions are irreversible is termed a voltage. In other words, emf, $\mathcal{E}$, is the maximum possible voltage that a galvanic cell can produce.

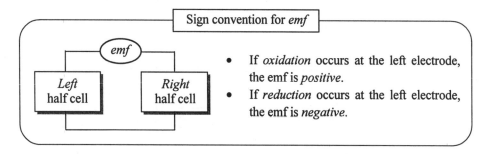

Electrical work (w_{elec}) per mole of reactant of a galvanic cell is the product of the charge transported (q, coulombs/mol) and the voltage (V, volts).

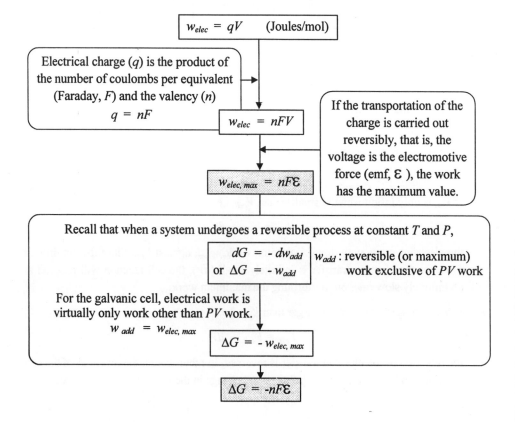

This equation is significant in that :

- It enables us to measure ΔG for the reaction by applying an opposing electric potential of magnitude $\mathcal{E}$ externally, which results in no current to flow.
- It means that when the chemical driving force (ΔG) is exactly balanced by the external opposing voltage ($-\mathcal{E}$), the whole system (the cell) is at equilibrium, i.e., electrochemical equilibrium.
- The measured ΔG can be used as a criterion of whether the process (or reaction) will take place spontaneously when all the species involved in the cell reactions are mixed in a reactor under otherwise same conditions.

Consider the general reaction

$$a\text{A} + b\text{B} = m\text{M} + n\text{N} \qquad \Delta G$$

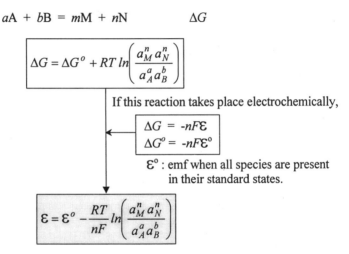

$$\Delta G = \Delta G^\circ + RT \ln\left(\frac{a_M^n a_N^n}{a_A^a a_B^b}\right)$$

If this reaction takes place electrochemically,

$$\Delta G = -nF\mathcal{E}$$
$$\Delta G^\circ = -nF\mathcal{E}^\circ$$

$\mathcal{E}^\circ$: emf when all species are present in their standard states.

$$\mathcal{E} = \mathcal{E}^\circ - \frac{RT}{nF} \ln\left(\frac{a_M^n a_N^n}{a_A^a a_B^b}\right)$$

This equation is known as the *Nernst equation*, and it plays a central role in electrochemistry.

- This equation enables us to determine how the emf of a cell should vary with composition.
- The emf at the standard states ($\mathcal{E}^\circ$) can be determined by constructing the cell with all the reagents at unit activity.
- Activities or activity coefficients of reagents can be obtained.

We now derive several important equations in terms of emf :

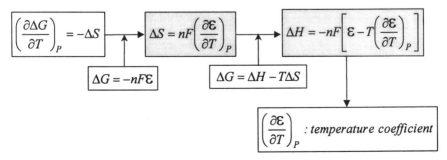

$$\left(\frac{\partial \Delta G}{\partial T}\right)_P = -\Delta S$$

$$\Delta S = nF\left(\frac{\partial \mathcal{E}}{\partial T}\right)_P$$

$$\Delta H = -nF\left[\mathcal{E} - T\left(\frac{\partial \mathcal{E}}{\partial T}\right)_P\right]$$

$$\Delta G = -nF\mathcal{E}$$

$$\Delta G = \Delta H - T\Delta S$$

$\left(\dfrac{\partial \mathcal{E}}{\partial T}\right)_P$: *temperature coefficient*

The value of ΔS is independent of temperature to a good approximation:

$$\Delta S = nF\left(\frac{\partial \varepsilon}{\partial T}\right)_P$$

Integration with ΔS being constant

$$\varepsilon_{T_2} = \varepsilon_{T_1} + \frac{\Delta S}{nF}(T_2 - T_1)$$

It can be seen in the above equation that the emf is a linear function of temperature. Thus, by measuring ε and the temperature coefficient, we can obtain the thermodynamic properties of the cell, ΔG, ΔS, ΔH.

Example 1

The figure shown below depicts the following redox reaction which occurs in an electrochemical manner:

$$Zn + CuSO_4 = Cu + ZnSO_4$$

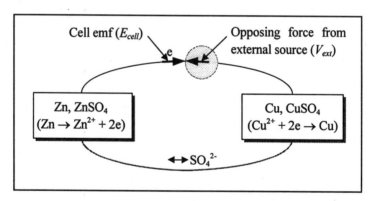

1) If $|\varepsilon_{cell}| > |V_{ext}|$, will Zn be oxidized or reduced?
2) If $|\varepsilon_{cell}| < |V_{ext}|$, In which direction is work done? On the cell or by the cell?
3) Find conditions for the cell to perform the maximum work.
4) The standard free energy change of the above reaction is -213,000 J at 25°C. If all species involved in the reaction are at their standard state, calculate the maximum amount of work the cell can perform per mole of Zn.
5) Calculate the standard emf ($\varepsilon°$) at 25°C.
6) If the cell consists of pure solid Zn and pure solid Cu,
   and CuSO$_4$(aqueous solution)
   and ZnSO$_4$(aqueous solution), both with the same activity,
   calculate the emf of the cell at 25°C.

1) The reaction will proceed from left to right with the net driving force of
   $\Delta V = |\mathcal{E}_{cell}| - |V_{ext}|$. For Zn, thus, Zn → Zn^{2+} : oxidation.

2) As the external voltage is larger than the cell emf, work is done on the system. This is an *electrolysis* process : i.e., current is consumed rather than produced.

3) The cell performs the maximum work when $|\mathcal{E}_{cell}| = |V_{ext}|$

4) As $\Delta G = -w_{max}$, $w_{max} = 213,000$ J/mol

5) $\Delta G° = -nF\mathcal{E}°$ $n = 2$, $F = 96,487$ coulombs, → $\mathcal{E}° = 1.104$ volts

6) Using the Nernst equation,

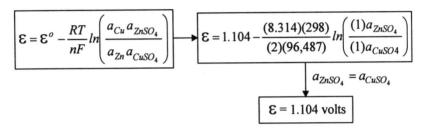

$$a_{ZnSO_4} = a_{CuSO_4}$$

$$\mathcal{E} = 1.104 \text{ volts}$$

Example 2

Is the following statement true?
"If the physical size of an electrochemical cell is doubled, the cell emf will also be doubled."

We check the validity of the statement with the following example :

$$Zn + CuSO_4 = Cu + ZnSO_4 \qquad \Delta G° = -213,000 \text{ J at } 298K$$

Suppose that
$a_{Zn} = 1$, $a_{Cu} = 1$,
$a_{ZnSO_4} = 50 a_{CuSO_4}$

$$\Delta G = \Delta G° + RT \ln\frac{a_{Cu} a_{ZnSO_4}}{a_{Zn} a_{CuSO_4}}$$

$$\Delta G = -213,000 + (8.314)(298)\ln\left(\frac{1 \times 50}{1 \times 1}\right) = -203,300 J$$

$$\Delta G = -nF\mathcal{E}$$
$$n = 2$$

$$\mathcal{E} = 1.054 \text{ volts}$$

If we double the stoichiometry of the reaction,

$$2Zn + 2CuSO_4 = 2Cu + 2ZnSO_4 \qquad \Delta G° = -426,000 \text{ J at } 298K$$

$$\Delta G = \Delta G^\circ + RT \ln\left(\frac{a_{Cu}^2 \, a_{ZnSO_4}^2}{a_{Zn}^2 \, a_{CuSO_4}^2}\right)$$

$a_{Zn} = 1,\ a_{Cu} = 1,\ a_{ZnSO_4} = 50 a_{CuSO_4}$

$$\Delta G = -426{,}000 + (8.314)(298)\ln\left(\frac{1\times 50^2}{1\times 1}\right) = -406{,}600\,J$$

$\Delta G = -nF\mathcal{E}$
$n = 4$

$\mathcal{E} = 1.054$ volts

As seen in the above, the electromotive force of a cell is an intensive property. Like temperature and pressure it is independent of the size of the system.

Exercises

1. The standard emf of the cell reaction

    $$Zn + CuSO_4 = Cu + ZnSO_4$$

    is 1.104 volts at 25°C. Calculate the equilibrium constant of the above reaction at 25°C.

2. The standard emf of the cell reaction

    $$Cd(s) + Hg_2Cl_2(s) = 2Hg(l) + CdCl_2(aq)$$

    is given as a function of temperature :

    $$\mathcal{E}^\circ = 0.487 + (13.3 \times 10^{-4}T - 2.4 \times 10^{-6}T^2,\ (\text{volts})$$

    Calculate the values of ΔG°, ΔH°, and ΔS° of the reaction at 35°C.

3. Calculate the emf at 25°C of the cell represented by the following reaction :

    $$Zn(s) + CdSO_4(aq) = Cd(s) + ZnSO_4(aq)$$

    The activities of $CdSO_4(aq)$ and $ZnSO_4(aq)$ are 7.0×10^{-3} and 3.9×10^{-3}, respectively, and cadmium is in the form of alloy with more noble metal ($a_{Cd} = 0.6$). The standard emf of the cell at 25°C is 0.36 volts.

5.2. Electrochemical Cells

5.2.1. Cells and Electrodes

The following figure gives a typical example of electrochemical cells (galvanic cells):

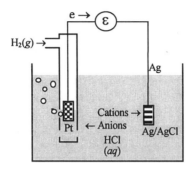

The cell consists of
- *a hydrogen electrode,*
- *a silver-silver chloride electrode, and*
- *an aqueous solution of HCl.*

The two electrodes are immersed in the HCl solution and connected to a potentiometer which measures the emf of the cell.

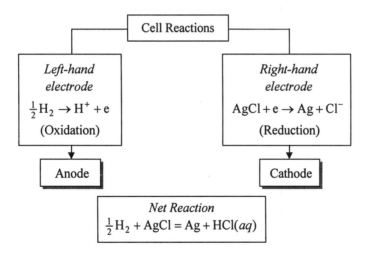

The convention is that,
- Since the left-hand electrode (anode) is in excess of electrons due to oxidation reaction, it is *negative*, and
- Since the right-hand electrode (cathode) is in deficit of electrons due to reduction reaction, it is *positive*.

It is awkward to present a cell in such a manner as described above. The cell assembly is thus, by convention, represented, without producing a figure like the one given above, by a diagram as shown below :

$$Pt \mid H_2(g, P = p \text{ atm}) \mid HCl(aq, a = m) \mid AgCl \mid Ag$$

or in abbreviation,

$$Pt \mid H_2 \mid HCl(aq) \mid AgCl \mid Ag$$

where P : pressure at p atm
 aq : aqueous solution
 a : activity of the value of m
 vertical line : phase boundary

When $P_{H_2} = 1\,atm$ and $a_{HCl} = 1$, the above cell drives electrons from the hydrogen electrode to the silver electrode with the emf of 0.222 V at 25°C. This means that oxidation occurs at the left-hand electrode, and reduction at the right-hand electrode. Therefore the emf of the cell is *positive* : $\varepsilon = + 0.222$ V.

If the cell were described in the opposite way, i.e.,

$$Ag \mid AgCl \mid HCl(aq) \mid H_2 \mid Pt$$

the emf would be *negative* ($\varepsilon = -0.222$ V) because reduction would occur at the left-hand electrode.

In summary, the convention of representation of the electrochemical cell is

> anode | solution | cathode

An important feature of electrochemical cells:
The reaction must be capable of being separated physically so that the direct chemical reaction is prevented from occurring.

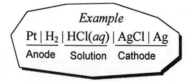

Reactants H_2 and AgCl are isolated at the separate electrodes, but maintain the electrical contact with each other through the aqueous HCl solution.

In some cases, two different solutions are used to prevent direct chemical reaction. A typical example is the *Daniell cell* illustrated below :

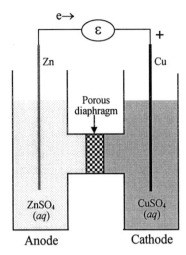

The cell comprises
- a zinc anode dipping into ZnSO$_4$ solution,
- a copper cathode dipping into CuSO$_4$ solution, and
- a porous diaphragm which prevents the two solutions from mixing, but allows electrical contact by the passage of SO_4^{2-} ions through it.

The two solutions constitute the *electrolyte* of the cell, which is the medium through which ionic current flows.

Cell reactions:

Anode: $Zn + SO_4^{2-}(aq) = ZnSO_4(aq) + 2e$ or $Zn = Zn^{2+}(aq) + 2e$
Cathode: $CuSO_4(aq) + 2e = Cu + SO_4^{2-}(aq)$ or $Cu^{2+}(aq) + 2e = Cu$
Net : $Zn + CuSO_4(aq) = ZnSO_4(aq) + Cu$ or $Zn + Cu^{2+}(aq) = Zn^{2+}(aq) + Cu$

Again, it is not convenient to draw the cell figure and write the cell reactions as shown above. The cell is thus conventionally represented by the following diagram :

$$Zn \,|\, ZnSO_4(aq) \vdots CuSO_4(aq) \,|\, Cu$$

or

$$Zn \,|\, Zn^{2+}(aq) \vdots Cu^{2+}(aq) \,|\, Cu$$

where the dashed vertical line represents the porous diaphragm separating the two aqueous solutions.

The emf of the cell depends on the activities of ZnSO$_4$ and CuSO$_4$ (recall the Nernst equation). Suppose all the reagents of the cell are in their standard states at 25°C, i.e., Zn (pure), Cu(pure), ZnSO$_4$(saturated) and CuSO$_4$(saturated).

$\boxed{\Delta G° = -nF\mathcal{E}°}$ $\Delta G° = -213,000$ J at 25°C → $\boxed{\mathcal{E}° = 1.104 \text{ V}}$
$n = 2$

Thus, $Zn \,|\, Zn^{2+}(aq) \vdots Cu^{2+}(aq) \,|\, Cu$ $\mathcal{E}° = 1.104$ V

If the reaction is written in the opposite direction, i.e.,

$$Cu \mid Cu^{2+}(aq) \vdots Zn^{2+}(aq) \mid Zn \qquad \varepsilon° = -1.104 \text{ V}$$

The emf of the cell at a non-standard state may be determined by the Nernst equation :

$$\varepsilon = \varepsilon^o - \frac{RT}{nF} \ln\left(\frac{a_{Cu} a_{ZnSO_4}}{a_{Zn} a_{CuSO_4}}\right)$$

The standard emf can be evaluated for all possible cells by one of the following two ways:
- Direct measurements of $\varepsilon°$ under standard conditions
- Measurements of ε under non-standard conditions and calculations of $\varepsilon°$ using the Nernst equation.

A complete list of the emf's of all possible cells, however, would be inordinately long and impractical. It would be much more convenient to develop some means of expressing the tendency of oxidation (or electron-donating power) of the individual electrodes. This can be done by using a *reference electrode* against which the electron-donating power of other electrodes is compared. It is agreed that the hydrogen electrode comprising H_2 gas at 1 atm and an aqueous solution containing hydrogen ions at unit activity (i.e., *standard hydrogen electrode*) is chosen as the reference electrode.

When we construct the following cell with all the reagents at their standard states, the standard emf of the cell at 25°C is found to be + 0.763 V :

$$Zn \mid Zn^{2+} \mid H^+ \mid H_2 \qquad \varepsilon° = +0.763 \text{ V}$$

Meaning of the *positive* emf
Oxidation occurs at the left-hand electrode.
$$Zn \rightarrow Zn^{2+} + 2e$$
Reduction occurs at the right-hand electrode.
$$2H^+ + 2e \rightarrow H_2$$

> In other words, Zn is stronger than H_2 in terms of electron-donating tendency.

When we construct the following cell with all the reagents at their standard states, the standard emf of the cell at 25°C is found to be - 0.337 V.

$$Cu \mid Cu^{2+} \mid H^+ \mid H_2 \qquad \varepsilon° = -0.337 \text{ V}$$

Meaning of the *negative* emf
Reduction occurs at the left-hand electrode.
$$Cu^{2+} + 2e \rightarrow Cu$$
Oxidation occurs at the right-hand electrode.
$$H_2 \rightarrow 2H^+ + 2e$$

> In other words, Cu is weaker than H_2 in terms of electron-donating tendency.

If the above results may be presented using a diagram as follows:

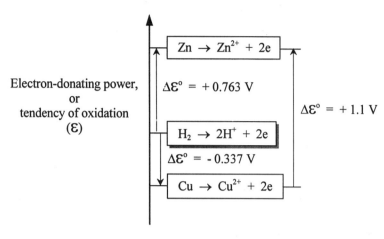

From the above figure one can easily see that Zn is stronger than Cu in terms of the electron-donating power and hence, if we construct a cell with Zn and Cu as follows, the standard emf of the cell will be + 1.1 V (= 0.763V + 0.337V).

$$Zn\,|\,Zn^{2+}(aq) \vdots Cu^{2+}(aq)\,|\,Cu \qquad \varepsilon° = 1.1\ V$$

Although it is impossible to measure the emf of a single half-cell or electrode like $H_2 \rightarrow 2H^+ + 2e$ and $Zn \rightarrow Zn^{2+} + 2e$, it is possible to assign an emf value for a cell, not an absolute value, but a value relative to the standard hydrogen electrode. This last statement is equivalent to saying that the standard hydrogen electrode is assigned a standard emf of zero. The value of the emf of a single electrode, or half-cell, which is relative to the standard hydrogen electrode, is called the *standard (single) electrode potential* or the *standard half-cell potential*.

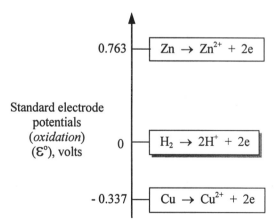

If we are concerned with the electron-accepting power, i.e., the tendency of reduction, the emf is the same in the numerical value as the one for the oxidation, but opposite in sign, i.e.,

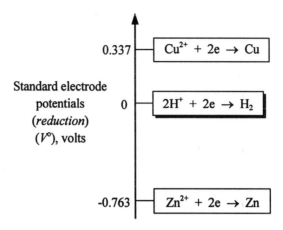

The standard oxidation and reduction potentials of a number of half-cells are given in Appendix III.

Example 1

Consider two hypothetical half-cells

$$A \rightarrow A^{a+} + ae^- \qquad \varepsilon_A^o$$
$$B \rightarrow B^{b+} + be^- \qquad \varepsilon_B^o$$

When a cell is built from these two half-cells, the cell reaction will be

$$bA + aB^{b+} \rightarrow bA^{a+} + aB$$

Prove the standard emf of the cell (ε^o) is given

$$\varepsilon^o = \varepsilon_A^o - \varepsilon_B^o$$

Recall that the emf is related to the free energy change by $\Delta G = -nF\varepsilon$.

$$\begin{array}{ll} (1) & A \rightarrow A^{a+} + ae^- \qquad \Delta G_A^o = -aF\varepsilon_A^o \\ (2) & B \rightarrow B^{b+} + be^- \qquad \Delta G_B^o = -bF\varepsilon_B^o \end{array}$$

$$\downarrow b \times (1) - a \times (2)$$

$$bA + aB^{b+} \rightarrow bA^{a+} + aB \qquad \Delta G^o = -(ab)F\varepsilon^o$$

When we add or subtract chemical equations, we also add and subtract changes in thermodynamic functions like U, H, S and G.

$$\boxed{\Delta G^o = b\Delta G_A^o - a\Delta G_B^o} \longrightarrow \boxed{\varepsilon^o = \varepsilon_A^o - \varepsilon_B^o}$$

$$\Delta G_A^o = -aF\varepsilon_A^o$$
$$\Delta G_B^o = -bF\varepsilon_B^o$$
$$\Delta G^o = -(ab)F\varepsilon^o$$

Half-cell potentials are directly combined *without* taking stoichiometric coefficients into consideration. However, calculation of free energy change *must take* stoichiometric coefficients into account. *Why?*

<u>There is a significant difference between ΔG and ε :</u>

- G is an *extensive property* of the system so that when the number of moles is changed, G must be adjusted accordingly.
- E is an *intensive property* of the system so that it is independent of the size of the system.

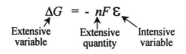

$$\Delta G = -nF\varepsilon$$

Extensive variable — Extensive quantity — Intensive variable

Example 2

Calculate $\varepsilon°$ and $\Delta G°$ for the cell at 25°C.

$$\text{Li} \,|\, \text{Li}^+(\text{aq}) \,|\, \text{Cd}^{2+}(\text{aq}) \,|\, \text{Cd}$$

(cathode / anode annotations)

From the table of the standard half-cell potentials

Li = Li$^+$ + e	$\varepsilon°_{Li}$ = 3.045 V
Cd = Cd^{2+} + 2e	$\varepsilon°_{Cd}$ = 0.403 V
Cell reaction 2Li + Cd^{2+} = 2Li$^+$ + Cd	$\varepsilon°$

1) Direct calculation from the standard half-cell emf's

$$\varepsilon° = \varepsilon°_{Li} - \varepsilon°_{Cd} = 3.045 - 0.403 = 2.642 V$$

2) Calculation from $\Delta G°$

$$\Delta G°_{Li} = -(1)(96{,}487)(3.045) = -293{,}800 J$$
$$\Delta G°_{Cd} = -(2)(96{,}487)(0.403) = -77{,}770 J$$
$$\Delta G° = 2\Delta G°_{Li} - \Delta G°_{Cd} = (2)(-293{,}800) - (-77{,}770) = -509{,}830 J$$

$$\boxed{\Delta G° = nF\varepsilon°} \xrightarrow{n=2} \boxed{\varepsilon° = 2.642 \text{ V}}$$

Example 3

Calculate the standard potential of the half-cell

$$\text{Cr} \rightarrow \text{Cr}^{2+} + 2e \qquad \varepsilon°$$

Given : (1) $\text{Cr} \rightarrow \text{Cr}^{3+} + 3e$ \qquad $\varepsilon°_{Cr/Cr^{3+}} = 0.74$ V

(2) $\text{Cr}^{2+} \rightarrow \text{Cr}^{3+} + e$ \qquad $\varepsilon°_{Cr^{2+}/Cr^{3+}} = 0.41$ V

$$\Delta G° = -nF\mathcal{E}°$$

$$\Delta G_i° = -nF\mathcal{E}_i°$$
$$\Delta G_1° = -(3)\times(96,487)\times(0.74) = -214,200 J$$
$$\Delta G_2° = -(1)\times(96,487)\times(0.41) = -39,560 J$$

$$\Delta G° = \Delta G_1° - \Delta G_2°$$
$$= -214,200 - (-39,560) = -174,640 J$$

$n = 2$

$$\mathcal{E}° = 0.905 \text{ V}$$

cf. Direct calculation : $\mathcal{E}° = \mathcal{E}°_{Cr/Cr^{3+}} - \mathcal{E}°_{Cr^{2+}/Cr^{3+}} = \underline{0.33V}$: *Incorrect*

"When half-cell potentials are combined to produce a new half-cell, the potentials are not additive. When in doubt, do the calculation of $\mathcal{E}$ from ΔG."

Exercises

1. Write the electrode reactions and the cell reactions for the following galvanic cells, and calculate the standard emf's of the cells at 25°C. Determine the positive electrodes :

   1) $Cu \mid CuCl_2(aq) \mid Cl_2(g)$, Pt

   2) Ag, $AgCl \mid HCl(aq) \parallel Hbr(aq) \mid AgBr$, Ag

2. Devise galvanic cells in which the cell reactions are the following :

   1) $Fe + CuSO_4(aq) = FeSO_4(aq) + Cu$

   2) $Pb(s) + PbO_2(s) + 2H_2SO_4(aq) = 2PbSO_4(s) + 2H_2O(l)$

   Indicate electrode reactions in each case. Calculate the standard emf of each of the cells at 25°C.

3. Calculate the standard emf ($\mathcal{E}°$) and the equilibrium constant (K) at 25°C for the reaction

$$Tl + Ag^+ = Tl^+ + Ag$$

5.2.2. Concentration Cells

When two half-cells are connected, an electrochemical cell is formed. The connection is made by bringing the solutions in the half-cells into contact so that ions can pass between them.
- If these two solutions are the same, there is no liquid junction, and we have *a cell without transference*.
- If these two solutions are different, transport of ions across the junction will cause irreversible changes in the two electrolytes, and we have *a cell with transference*.

The driving force of a cell may come from a chemical reaction or from a physical change. When the driving force of a cell is changes in composition (concentration) of species (or of gas pressures), the cell is called *concentration cell*. The change in concentration can occur either in the electrolyte or in the electrodes.

The electrochemical cells can therefore be classified as follows :

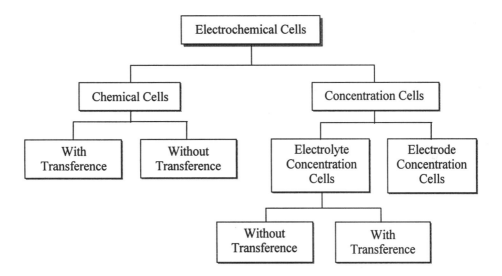

To form an electrode concentration cell the electrode material must have a variable concentration. Amalgam and gaseous electrodes fall into this classification. An example of electrode concentration cells is the one in which two amalgam electrodes of different concentrations dip into a solution containing the solute metal ions.

$$\text{Hg-Cd}(a_{Cd(1)}) \mid \text{CdSO}_4(\text{solution}) \mid \text{Hg-Cd}(a_{Cd(2)})$$

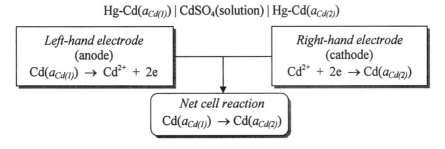

$$\varepsilon = \varepsilon^\circ - \frac{RT}{nF} \ln\left(\frac{a_{Cd(2)}}{a_{Cd(1)}}\right)$$

$n = 2$
$\varepsilon^\circ = 0$
(because at the standard state the cell reactions at both sides are the same)

$$\varepsilon = -\frac{RT}{2F} \ln\left(\frac{a_{Cd(2)}}{a_{Cd(1)}}\right)$$

As seen above, no chemical change occurs, and the process consists of the transfer of cadmium from an amalgam of one concentration to that of another concentration. The cadmium will tend to go spontaneously from the high activity amalgam to that of low activity.

- If $a_{Cd(1)} > a_{Cd(2)}$, $\varepsilon > 0$ and hence the reaction proceeds as indicated in the cell reaction.
- If $a_{Cd(1)} < a_{Cd(2)}$, $\varepsilon < 0$ and hence the reaction proceeds in the opposite direction.

If the activity of cadmium in one amalgam is known, the activity of cadmium in the other amalgam can be determined by measuring the emf of the cell.

An electrode concentration cell can be constructed using two electrodes which consist of a gas at different partial pressures. The following is an example of the hydrogen electrode concentration cell:

$$\text{Pt} \mid H_2(P_1) \mid HCl(aq) \mid H_2(P_2) \mid \text{Pt}$$

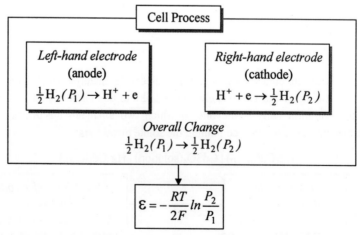

The emf of this cell is thus determined by the hydrogen pressures.

It has been found that $ZrO_2(s)$ stabilised with CaO or $YO_{1.5}$ is an ionic conductor of oxygen ions in certain ranges of oxygen pressure and temperature. Using this material as the *solid electrolyte*, an oxygen concentration cell can be constructed:

$$Pt, O_2(g, P_{O_2(a)}) \mid ZrO_2+CaO \mid O_2(g, P_{O_2(c)}), Pt$$

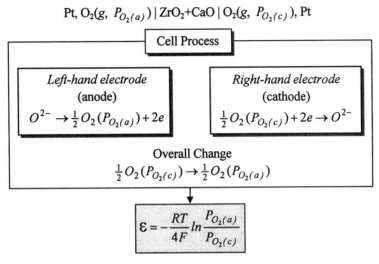

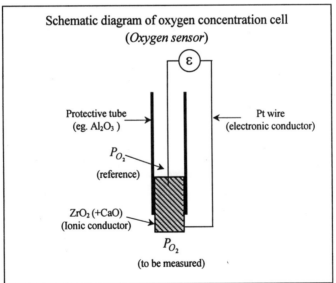

This type of oxygen concentration cell called the *oxygen sensor* is of considerable use in high temperature processing of metals and materials and in combustion processes. The oxygen sensor with proper design enables *in situ* measurements of
- oxygen partial pressure in a gas mixture,
- equilibrium oxygen potential in a metal-metal oxide mixture, and
- oxygen content in a liquid metal.
- activity of a metal in an alloy
- activity of an oxide in an oxide solution.

The oxygen pressure of the reference electrode ($P_{O_2(ref)}$) can be fixed either by using a gas mixture containing oxygen of known pressure (e.g., air), or by using a metal-metal oxide mixture.

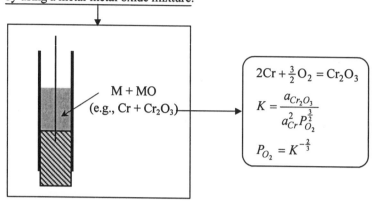

Suppose a cell is constructed by using metal-metal oxide couples for both electrodes:

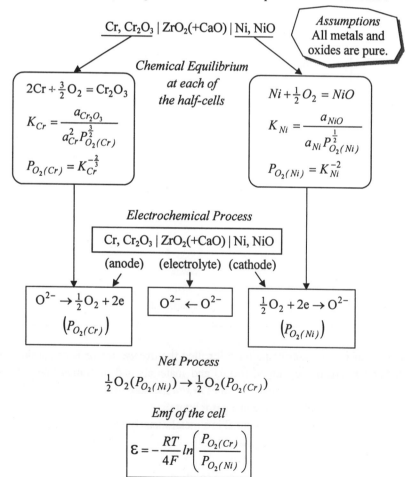

Provided that the oxygen pressure of the reference electrode is known (say, $P_{O_2(Cr)}$), the oxygen pressure of the other electrode ($P_{O_2(Ni)}$) can be determined by measuring the emf of the cell. This enables the determination of the free energy of formation of the oxide (say NiO).

A cell in which the emf is derived only from the free energy change of dilution of the electrolyte is called the *electrolyte concentration cell*.
Consider a simple cell

$$Pt \mid H_2 \mid HCl(aq) \mid AgCl \mid Ag$$

The net cell reaction is

$$\tfrac{1}{2}H_2 + AgCl \rightarrow Ag + HCl(aq)$$

If two such cells are electrically connected in the opposed manner, the combination constitutes a cell that may be written

$$Ag \mid AgCl \mid HCl(aq, a_1) \mid H_2 \mid HCl(aq, a_2) \mid AgCl \mid Ag$$

The overall change in this cell is simply the difference between the changes in the two separate cells.

Cell Reactions

Left-hand electrode	Right-hand electrode
$HCl(aq, a_1) + Ag \rightarrow AgCl + \tfrac{1}{2}H_2$	$AgCl + \tfrac{1}{2}H_2 \rightarrow HCl(aq, a_2) + Ag$

Net Reactions

$$HCl(aq, a_1) \rightarrow HCl(aq, a_2)$$

Note that there is no direct transference of the electrolyte (HCl) from one side to the other. HCl is removed from the left-hand side by the left-hand electrode reaction and it is added to the right-hand side by the right-hand electrode reaction. This cell is an example of a *electrolyte concentration cell without transference*.

If two electrolytes of different concentrations are directly in contact with each other, this give rise to a *junction potential*. An example of a concentration cell with a liquid junction is

$$Pt \mid H_2\,(1\,atm) \mid HCl\,(aq, c_1) \vdots HCl\,(aq, c_2) \mid H_2\,(1\,atm) \mid Pt$$

Cell Reaction

Anode reaction	Cathode reaction
$\tfrac{1}{2}H_2(1\,atm) \rightarrow H^+(c_1) + e$	$H^+(c_2) + e \rightarrow \tfrac{1}{2}H_2(1\,atm)$

Net reaction

$$H^+(c_2) \rightarrow H^+(c_1)$$

In the above cell, the two HCl solutions are in electrolytic contact, but are prevented from mechanical mixing. In general, this is done by means of a porous diaphragm or by stiffening one of the solutions at its point of contact with the other by agar-agar or gelatine. Since the direct contact between solutions of different concentrations is not a balanced state, as required for reversible processes, the system is not directly susceptible to thermodynamic analysis. The liquid junction problem may be circumvented by connecting the different solutions by means of a bridge containing a saturated KCl solution.

Example 1

Concentration cells have been used extensively to determine thermodynamic properties of metallic solutions. The following cell is built to measure the thermodynamic properties of Mg in Mg-M alloys :

$$Mg(l, \text{ pure}) \,|\, MgCl_2, CaCl_2 \,|\, Mg\text{-}M(l)$$

Choose the correct one in the following :

1) The alloying element M should be *more* noble than Mg.
2) The alloying element M should be *less* noble than Mg.

The element M has to be nobler than M. In other words, the chloride of M should be much less stable than either $MgCl_2$ or $CaCl_2$. Otherwise the element M would react with $MgCl_2$ or $CaCl_2$ in preference of Mg.

The cell process of the above cell is

$$Mg(l) \rightarrow Mg(\text{in alloy})$$

$$\varepsilon = -\frac{RT}{2F} \ln\left(\frac{a_{Mg(alloy)}}{a_{Mg}}\right)$$

$$a_{Mg} = 1$$

$$a_{Mg(alloy)} = \exp\left(-\frac{2F\varepsilon}{RT}\right)$$

$$\overline{G}_{Mg}^M = RT \ln a_{Mg(alloy)}$$

Thus an electrochemical measurement enables the determination of the activity of a component in the solution and hence the partial molar free energy of the component. If we measure the emf in a range of temperature, we can determine $\overline{S}_{Mg}^M$ and $\overline{H}_{Mg}^M$ from the relationships developed in the previous section :

$$\overline{S}_{Mg}^M = 2F\left(\frac{\partial \varepsilon}{\partial T}\right)_P \qquad \overline{H}_{Mg}^M = -2F\varepsilon + 2FT\left(\frac{\partial \varepsilon}{\partial T}\right)_P$$

Example 2

The standard free energy of formation of an oxide can be determined by measuring the emf of a cell appropriately designed. Consider the following cell :

$$M, M_aO_b | ZrO_2(+CaO) | O_2(g, P_{O_2(c)}) \qquad \varepsilon$$

Prove that the free energy of formation of M_aO_b, $\Delta G^o_{f,M_aO_b}$, is given by the equation

$$\boxed{\Delta G^o_{f,M_aO_b} = -4\left(\frac{b}{2}\right)F\varepsilon + \left(\frac{b}{2}\right)RT \ln P_{O_2(c)}}$$

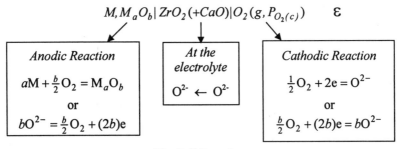

$$\boxed{\Delta G^o_{f,M_aO_b} = -4\left(\frac{b}{2}\right)F\varepsilon + \left(\frac{b}{2}\right)RT \ln P_{O_2(c)}}$$

If a metal-metal oxide couple (e.g., Me-MeO) is used for the reference electrode (the cathode in the present example), $P_{O_2(c)}$ must be replaced with the equilibrium oxygen pressure of the couple :

$$Me + \tfrac{1}{2}O_2 = MeO \qquad K = \frac{a_{MeO}}{a_{Me} P^{\frac{1}{2}}_{O_2(c)}} \rightarrow P_{O_2(c)} = K^{-2}$$

Example 3

Galvanic cells with solid ZrO_2 (+CaO) electrolyte are used as the *oxygen sensor* for high temperature applications. The oxygen sensor enables determination of the activity of a metal element in an alloy. Consider the following cell:

$$Ni, NiO \mid ZrO_2 (+CaO) \mid \underline{M} \text{ (in alloy)} \mid MO$$

Prove that

$$\varepsilon = \varepsilon^o - \frac{RT}{2F} \ln a_M$$

The cell reaction can be written as

$$Ni + MO = NiO + M$$

From the Nernst equation

$$\varepsilon = \varepsilon^o - \frac{RT}{2F} \ln\left(\frac{a_M \, a_{NiO}}{a_{Ni} \, a_{MO}}\right)$$

$a_{NiO} = 1, \quad a_{MO} = 1, \quad a_{Ni} = 1$

$$\varepsilon = \varepsilon^o - \frac{RT}{2F} \ln a_M$$

An important point: For the analysis described above to be valid, all other components in the metal alloy must be considerably nobler than M so that these are practically inert under the oxygen potential prevailing in the electrode.

Example 4

The method described in the example 3 can also be used to determine the activity of an oxide in the oxide solution. Consider the following cell:

$$M(l) \mid MO(l) \mid ZrO_2 (+CaO) \mid Me(l) \mid MeO(l, \text{ in solution})$$

Prove that

$$\varepsilon = \varepsilon^o + \frac{RT}{2F} \ln a_{MeO}$$

Cell Reaction
$$M + MeO = Me + MO$$

$$\varepsilon = \varepsilon^o - \frac{RT}{2F} \ln\left(\frac{a_{Me} \, a_{MO}}{a_M \, a_{MeO}}\right) \quad a_M = 1, \quad a_{MO} = 1, \quad a_{Me} = 1 \quad \longrightarrow \quad \varepsilon = \varepsilon^o + \frac{RT}{2F} \ln a_{MeO}$$

Example 5

The oxygen sensor is capable of in situ measurement of oxygen potential in liquid metals. Consider the following cell :

$$M, MO \mid ZrO_2 (+CaO) \mid \underline{O} \text{ (dissolved in metal)}$$

Prove that

$$\varepsilon = \varepsilon^o + \frac{RT}{2F} \ln a_O$$

Cell Reaction

$$M + \underline{O} = MO$$

$$\varepsilon = \varepsilon^o - \frac{RT}{2F} \ln\left(\frac{a_{MO}}{a_M \, a_O}\right)$$

$$a_M = 1, \quad a_{MO} = 1$$

$$\varepsilon = \varepsilon^o + \frac{RT}{2F} \ln a_O$$

where ε^o is the standard emf of the cell with the unit oxygen pressure at the cathode:

$$M, MO \mid ZrO_2 (+CaO) \mid O_2 \, (P_{O_2} = 1 atm)$$

Exercises

1. The emf of the following cell

$$Mg(l) \mid MgCl_2\text{-}CaCl_2(l) \mid Mg_2Si(s), Si(s)$$

   was found to be

$$\varepsilon^o = 0.21767 - 8.607 \times 10^{-5} T, \quad (V)$$

   Express the standard free energy of formation of $Mg_2Si(s)$ as a function of temperature.

2. The emf of the following cell was measured at a number of different temperatures :

$$Cu_2O(s), CuO(s) \mid ZrO_2(+CaO) \mid O_2(g, air)$$

   The results were reported as follows :

T (K)	973	1023	1073	1123	1173	1223	1273
ε (mV)	170.0	143.9	117.7	91.6	65.5	39.3	13.2

Calculate the standard free energy of formation of CuO(s) as a function of temperature. The free energy of formation of Cu_2O is given :

$$\Delta G^o_{f,Cu_2O} = -168,400 + 71.25T, \quad \text{J/mole}$$

3. The activity of chromium in liquid Ni-Cr alloys was measured using the following cell :

$$Pt \mid Cr(s), Cr_2O_3(s) \mid ZrO_2(+CaO) \mid Ni\text{-}Cr(l), Cr_2O_3 \mid Pt$$

The cell emf was measured to be 125 mV at 1,600°C for the Ni-Cr alloy of N_{Cr} = 0.109. Calculate the activity of Cr in the alloy.

4. The oxygen content of molten iron can be measured using the following cell :

$$Pt \mid Cr(s), Cr_2O_3(s) \mid ZrO_2(+CaO) \mid \underline{O} \text{ (in liquid Fe)} \mid Mo$$

Derive an equation which relates the measured emf to the oxygen content (wt%) in the Fe melt at 1,600°C. The following data are given :

$$2Cr + \tfrac{3}{2}O_2 = Cr_2O_3 \quad \Delta G^o_{Cr} = -1,110,100 + 247.32T, \quad \text{Jmole}^{-1}$$

$$\tfrac{1}{2}O_2(g) = O(1\text{wt\%}) \quad \Delta G^o_O = -117,150 - 2.887T, \quad \text{Jmole}^{-1}$$

1 wt% in Fe melt standard state.

5.3. Aqueous Solutions

5.3.1. Activities in Aqueous Solutions

The thermodynamics properties of an electrolytic solution are generally described by using the activities of different ionic species present in the solution. The problem of defining activities is however somewhat more complicated in electrolytic solution than in solutions of nonelectrolytes. The requirement of overall electrical neutrality in the solution prevents any increase in the charge due to negative ions. Consider the 1:1 electrolyte AB which dissociates into A^+ ions and B^- ions in the aqueous solution.

$$AB = A^+ + B^-$$

The partial molar free energies of the two ions, $\overline{G}_{A^+}$, $\overline{G}_{B^+}$, are,

$$\overline{G}_{A^+} = G^o_{A^+} + RT \ln a_{A^+} \quad \text{and} \quad \overline{G}_{B^-} = G^o_{B^-} + RT \ln a_{B^-}$$

From the definition of the partial molar free energy,

$$\overline{G}_{A^+} = \left(\frac{\partial G}{\partial m_{A^+}} \right)_{m_{B^-},T,P}$$

$$\overline{G}_{B^-} = \left(\frac{\partial G}{\partial m_{B^-}}\right)_{m_{A^+}, T, P}$$

Where m_{A^+} and m_{B^-} are the molalities of A^+ and B^- ions, respectively, in the solution. The *molality* is defined as the number of moles of solute in 1,000g of water (H_2O).

As the molality of an ion (m_{A^+} and m_{B^-}) cannot be altered independently, it is not possible to measure either $\overline{G}_{A^+}$ or $\overline{G}_{B^+}$. In order to overcome this difficulty, we introduce *mean thermodynamics properties* of two ions.

Suppose that n moles of AB are dissolved in water :

$$nAB = nA^+ + nB^-$$

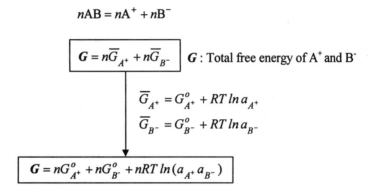

G in the above equation is the total free energy of $2n$ moles of ions (n moles of A^+ ion and n moles of B^- ions). If we define the mean partial molar free energy of ion ($\overline{G}_\pm$) as

$$\overline{G}_\pm = \frac{G}{(n_+ + n_-)}$$

where n_+ is the number of moles of positive ions and n_- is the number of moles of negative ions.

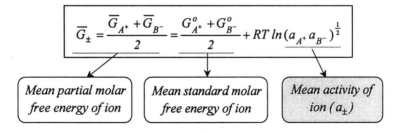

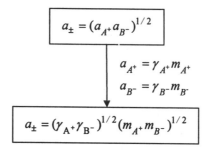

Activities in aqueous solution are generally based on the *1 molality standard state*.

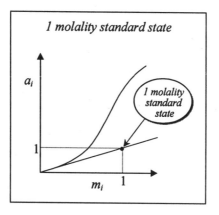

The *mean ionic molality* ($m_\pm$) and the *mean ionic activity coefficient* ($\gamma_\pm$) are defined as

$$m_\pm = (m_{A^+} m_{B^-})^{1/2} \qquad \gamma_\pm = (\gamma_{A^+} \gamma_{B^-})^{1/2}$$

$$a_\pm = \gamma_\pm m_\pm$$

As the activities in aqueous electrolyte solutions are defined with respect to the 1 molality standard state (or infinitely dilute solution standard state), the activity of an ionic species becomes equal to its molality as the concentration approaches zero (Henry's Law).

$$\lim_{m_{A^+} \to 0} \left(\frac{a_{A^+}}{m_{A^+}} \right) = 1 \qquad \lim_{m_{B^-} \to 0} \left(\frac{a_{B^-}}{m_{B^-}} \right) = 1$$

Thus, at infinite dilution, $\gamma_{A^+} = 1$, $\gamma_{B^-} = 1$, and $\gamma_\pm = 1$, and hence

$$a_\pm = m_\pm$$

For a 1-1 electrolyte, $m_{A^+} = m_{B^-} = m^o_{AB}$, and hence $a_\pm = m^o_{AB}$.

We now generalise our discussion with more complex types of electrolyte. Consider the nonsymmetrical electrolyte, A_xB_y, which dissociates into A^{z+} positive ions and B^{z-} negative ions in an aqueous solution:

$$A_xB_y = xA^{z+} + yB^{z-}$$

For the dissolution of one mole of A_xB_y

$$\boxed{G = x\overline{G}_{A^+} + y\overline{G}_{B^-}}$$

$$\overline{G}_{A^{z+}} = G^o_{A^{z+}} + RT \ln a_{A^{z+}}$$
$$\overline{G}_{B^{z-}} = G^o_{B^{z-}} + RT \ln a_{B^{z-}}$$

$$\boxed{G = xG^o_{A^{z+}} + yG^o_{B^{z-}} + RT \ln(a^x_{A^{z+}} a^y_{B^{z-}})}$$

G in the above equation is the total free energy of $(x+y)$ moles of ions (x moles of A^{z+} ion and y moles of B^{z-} ions). If we define the mean partial molar free energy of ion ($\overline{G}_\pm$) as

$$\boxed{\overline{G}_\pm = \frac{G}{(n_+ + n_-)}}$$

$$n_+ = x, \quad n_- = y$$

$$\boxed{\overline{G}_\pm = \frac{G}{x+y} = \frac{x\overline{G}_{A^{z+}} + y\overline{G}_{B^{z-}}}{x+y} = \frac{xG^o_{A^+} + yG^o_{B^-}}{x+y} + RT \ln(a^x_{A^{z+}} a^y_{B^{z-}})^{\frac{1}{x+y}}}$$

$$\boxed{a_\pm = (a^x_{A^{z+}} a^y_{B^{z-}})^{\frac{1}{x+y}}}$$

$$a_{A^{z+}} = \gamma_{A^{z+}} m_{A^{z+}}$$
$$a_{B^{z-}} = \gamma_{B^{z-}} m_{B^{z-}}$$

$$a_\pm = (\gamma^x_{A^{z+}} \gamma^y_{B^{z-}})^{\frac{1}{x+y}} (m^x_{A^{z+}} m^y_{B^{z-}})^{\frac{1}{x+y}}$$

$$m_\pm = (m^x_{A^{z+}} m^y_{B^{z-}})^{\frac{1}{x+y}}$$

$$\gamma_\pm = (\gamma^x_{A^{z+}} \gamma^y_{B^{z-}})^{\frac{1}{x+y}}$$

$$\boxed{a_\pm = \gamma_\pm m_\pm}$$

The total Gibbs free energy of the ions in the electrically neutral solution is the sum of the free energies of positive ions and of negative ions :

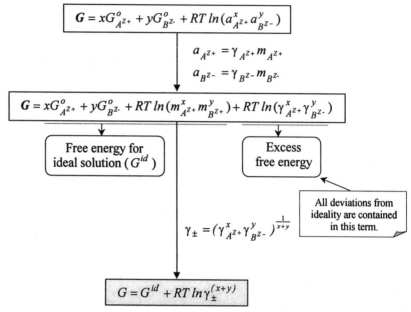

Observed deviations from ideal behaviour are ascribed to electrical interactions between ions. Oppositely charged ions attract each other. As a result, in the immediate neighbourhood of a given ion, an oppositely charged ion is more likely to be found. Overall the solution is electrically neutral, but near any given ion there is an excess of oppositely charged ions. Consequently the chemical potential of an ion is lowered as a results of its electrostatic interaction with its ionic neighbours. This lowering of energy appears as the difference between the Gibbs free energy G and the ideal value G^{id} of the solution; i.e., $RT \ln \gamma_{\pm}^{(x+y)}$.

On the assumption that deviations of a dilute solution from ideality are caused entirely by the electrostatic interactions, the activity coefficient can be calculated from the *Debye-Hückel limiting law* :

$$\log \gamma_{\pm} = -0.509 |z_+ z_-| I^{1/2}$$

where

z_+, z_- : charge numbers of positive and negative ions, respectively
I : ionic strength of the solution

$$I = \frac{1}{2} \sum_i z_i^2 m_i$$

Example 1

Consider the solution of $La_2(SO_4)_3$ at a molality m. Express the mean ionic activity $a_\pm$ in terms of $\gamma_\pm$ and m.

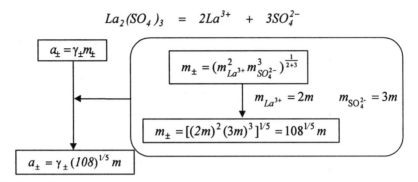

$$La_2(SO_4)_3 = 2La^{3+} + 3SO_4^{2-}$$

$$a_\pm = \gamma_\pm m_\pm$$

$$m_\pm = (m_{La^{3+}}^2 \cdot m_{SO_4^{2-}}^3)^{\frac{1}{2+3}}$$

$$m_{La^{3+}} = 2m \qquad m_{SO_4^{2-}} = 3m$$

$$m_\pm = [(2m)^2 (3m)^3]^{1/5} = 108^{1/5} m$$

$$a_\pm = \gamma_\pm (108)^{1/5} m$$

Example 2

Values of the mean ionic activity coefficients for several electrolytes in water at 25°C are given in the following table :

Mean activity coefficients $\gamma_\pm$ for strong electrolytes at 25°C

Electro- lytes	Molality(m)									
	0.001	0.002	0.005	0.01	0.05	0.1	0.5	1.0	2.0	4.0
HCl	0.996	0.952	0.928	0.904	0.830	0.796	0.758	0.809	1.01	1.76
HNO_3	0.965	0.951	0.927	0.902	0.823	0.785	0.715	0.720	0.783	0.982
H_2SO_4	0.830	0.757	0.639	0.544	0.340	0.265	0.154	0.130	0.124	0.171
$CaCl_2$	0.89	0.85	0.785	0.725	0.57	0.515	0.52	0.71		
$CuCl_2$	0.89	0.85	0.78	0.72	0.58	0.52	0.42	0.43	0.51	
$CuSO_4$	0.74		0.53	0.41	0.21	0.16	0.068	0.047		
$FeCl_2$	0.89	0.86	0.80	0.75	0.62	0.58	0.59	0.67		
KCl	0.965	0.952	0.927	0.901	0.815	0.769	0.651	0.606	0.576	0.579
K_2SO_4	0.89		0.78	0.71	0.52	0.43				
$MgCl_2$							0.56	0.52	0.62	1.05
NaCl	0.966	0.953	0.929	0.904	0.823	0.780	0.68	0.66	0.67	0.78
$PbCl_2$	0.86	0.80	0.70	0.61						
$ZnCl_2$	0.88	0.84	0.77	0.71	0.56	0.50	0.38	0.33		
NaOH						0.82		0.69	0.68	
KOH				0.92	0.90	0.82	0.8	0.73	0.76	
NaBr	0.966			0.934	0.914	0.844	0.800	0.695	0.686	
$MgSO_4$					0.40	0.22	0.18	0.088	0.064	
$ZnSO_4$	0.70	0.61	0.48	0.39		0.15	0.065	0.045	0.036	

Calculate the mean activities of HCl and $CuCl_2$ in the aqueous solution at 25°C. The concentration of both electrolytes is 0.01 molality.

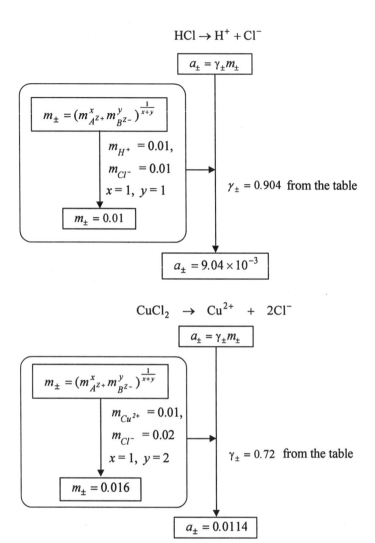

Example 3

Estimate the mean activity coefficient of the aqueous solution of KCl at 0.005 molal at 25°C.

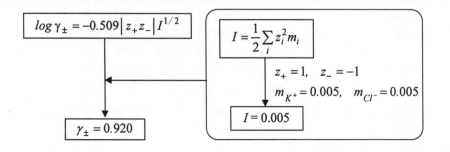

Example 4

The solubility of AgCl in water is 1.274×10^{-5} molal at 25°C. Calculate the standard Gibbs free energy change of reaction

$$AgCl(s) \rightarrow Ag^+(aq) + Cl^-(aq)$$

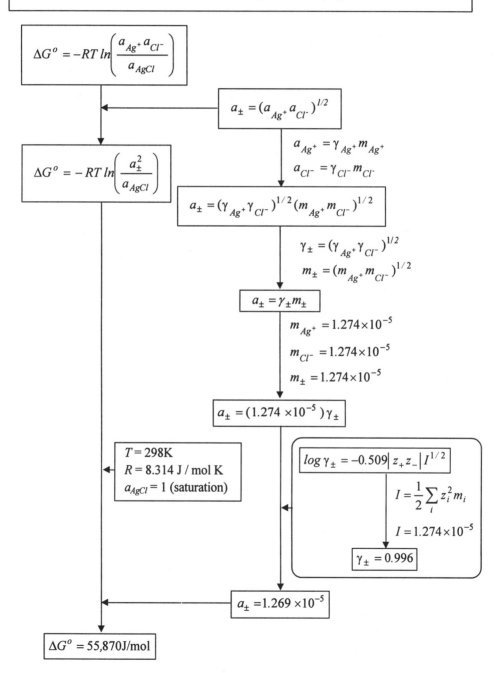

5.3.2. Solubility Products

Consider the dissolution of a sparingly soluble salt or electrolyte A_xB_y. Saturation of the aqueous solution occurs when A_xB_y has dissolved to the extent that the activity of A_xB_y in the solution, with the respect to the solid A_xB_y as the standard state, is unity.

$$A_xB_y = xA^{z+}(aq) + yB^{z-}(aq)$$

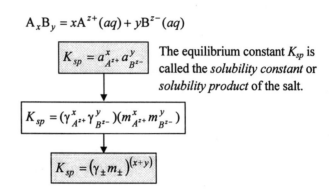

$$\boxed{K_{sp} = a_{A^{z+}}^x \, a_{B^{z-}}^y}$$ The equilibrium constant K_{sp} is called the *solubility constant* or *solubility product* of the salt.

$$\boxed{K_{sp} = (\gamma_{A^{z+}}^x \gamma_{B^{z-}}^y)(m_{A^{z+}}^x m_{B^{z-}}^y)}$$

$$\boxed{K_{sp} = (\gamma_\pm m_\pm)^{(x+y)}}$$

Example 1

Evaluate the solubility constant K_{sp} of NaCl at 25°C using cell potential data and the free energy of formation of NaCl:

$$2Cl^- = Cl_2 + 2e \qquad \varepsilon° = -1.360V$$
$$Na = Na^+ + e \qquad \varepsilon° = 2.714V$$
$$Na(s) + 1/2Cl_2(g) = NaCl(s) \qquad \Delta G° = -383{,}880 \text{ J mol}^{-1} \text{ at 298K}$$

---

1) $Cl^- = 1/2Cl_2 + e \qquad \Delta G°_{Cl} = -nF\varepsilon° = -(1)(96487)(-1.360) = 131{,}220$ J
2) $Na = Na^+ + e \qquad \Delta G°_{Na} = -nF\varepsilon° = -(1)(96487)(2.714) = -261{,}870$ J
3) $Na(s) + 1/2Cl_2(g) = NaCl(s) \qquad \Delta G°_{NaCl} = -383{,}880$ J

Combination of the above reactions yields

$$NaCl(s) = Na^+(aq) + Cl^-(aq) \qquad \Delta G° = -9{,}210 \text{ J}$$

$$K_{sp} = (\gamma_\pm m_\pm)^2 = \exp\left(\frac{-\Delta G°}{RT}\right) = 41.2$$

For NaCl,

$$m_\pm = m_{Na^+} = m_{Cl^-} = m_{NaCl}.$$

$$(\gamma_\pm m_\pm)^2 = (\gamma_\pm m_{NaCl})^2 = 41.2$$

or

$$\gamma_\pm m_{NaCl} = 6.42$$

Example 2

Calculate the concentrations of H^+ and OH^- in water at 25°C. Use cell potential data.

From the table of the standard half-cell potentials,

$$H_2 + 2OH^- = 2H_2O + 2e \qquad \mathcal{E}° = 0.828V$$
$$H_2 = 2H^+ + 2e \qquad \mathcal{E}° = 0 V$$

Combination yields

$$H_2O = H^+ + OH^- \qquad \mathcal{E}° = -0.828V$$

$$K_{sp} = \frac{a_{H^+} a_{OH^-}}{a_{H_2O}} = \exp\left(-\frac{\Delta G°}{RT}\right) = \exp\left(-\frac{-nF\mathcal{E}°}{RT}\right) = 9.9 \times 10^{-15}$$

Since

$$a_{H^+} = \gamma_{H^+} m_{H^+}, \quad a_{OH^-} = \gamma_{OH^-} m_{OH^-}, \quad a_{H_2O} = 1, \quad \gamma_{\pm} = (\gamma_{H^+} \gamma_{OH^-})^{\frac{1}{2}}$$

$$\boxed{\gamma_{\pm}^2 m_{H^+} m_{OH^-} = 9.9 \times 10^{-15}}$$

$\gamma_{\pm} \cong 1$ (very dilute)
$m_{H^+} = m_{OH^-}$ (from stoichiometry)

$$\boxed{m_{H^+} = 10^{-7}}$$

Definition
pH = -log[H^+]
where [H^+] : molarity of H^+

Molarity:
Number of moles of solute per liter of water.

For dilute solutions,
molarity $\cong$ molality

$$\boxed{pH = 7 \text{ for water}}$$

pH
← lower   7   higher →
acidic   basic
        neutral

> **Example 3**
>
> The thermodynamic behaviour of weak electrolytes is based on the conditions for equilibrium between dissociated ions and undissociated portion in the solution. Prove that, for most weak electrolytes, the degree of dissociation increases as the electrolyte concentration decreases.

Consider the weak electrolyte A_xB_y

$$A_xB_y(aq) = xA^{z+}(aq) + yB^{z-}(aq)$$

$$K_C = \frac{a_{A^{z+}}^x \, a_{B^{z-}}^y}{a_{A_xB_y}}$$

$a_i = \gamma_i m_i$

$$K_C = \left(\frac{\gamma_{A^{z+}}^x \, \gamma_{B^{z-}}^y}{\gamma_{A_xB_y}}\right)\left(\frac{m_{A^{z+}}^x \, m_{B^{z-}}^y}{m_{A_xB_y}}\right)$$

In dilute solutions, the activity coefficients, γ_i, approach 1. Thus

$$\left(\frac{\gamma_{A^{z+}}^x \, \gamma_{B^{z-}}^y}{\gamma_{A_xB_y}}\right) \cong 1$$

$$K_C = \frac{m_{A^{z+}}^x \, m_{B^{z-}}^y}{m_{A_xB_y}}$$

Let
m : molality of the electrolyte dissolved
ξ : fraction of dissociation of the dissolved electrolyte

$m_{A^{z+}} = x\xi m$,
$m_{B^{z-}} = y\xi m$,
$m_{A_xB_y} = (1-\xi)m$

$$K_C = \frac{(x^x y^y)\xi^{(x+y)} m^{(x+y-1)}}{(1-\xi)}$$

For dilute solutions, $\xi \ll 1$.

$$\xi = \left(\frac{K_C}{x^x y^y}\right)^{\frac{1}{x+y}} m^{(\frac{1}{x+y}-1)}$$

As m increases, $m^{(\frac{1}{x+y}-1)}$ decreases, and hence ξ decreases.

Example 4

A strong electrolyte is defined as a compound that, when added to a solvent, dissociates completely into its ionic components. If a strong acid such as HCl is dissolved in water, it dissociates completely into H^+ ions and Cl^- ions. Calculate the pH of a 0.1 molal solution of HCl in water.

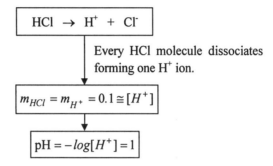

Example 5

A solution is prepared by mixing two litres of 0.02 molal HCl and one litre of 0.05 molal NaOH. Calculate the pH of the solution. HCl and NaOH are both strong electrolytes.

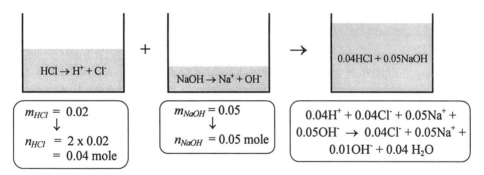

Upon mixing, 0.04 mole of H^+ reacts with 0.04 mole of OH^- to form 0.04 mole of H_2O, leaving 0.01 mole of OH^-. The total volume of the solution becomes 3 litres.

$$m_{OH^-} \cong [OH^-] = \frac{0.01}{3} = 0.0033 \, mole/litre$$

$$[H^+][OH^-] = 10^{-14}$$

$$[H^+] = 3.03 \times 10^{-12} \, mole/litre$$

$$pH = 11.5$$

Exercises

1. The solubility product of Cu_2S is 3×10^{-48}. Calculate the solubility of this salt.

2. 5 grams of $Pb(NO_3)_2$ is added to 1 litre of 0.01molal NaCl solution. Would $PbCl_2$ precipitate? The solubility product for $PbCl_2$ is 2×10^{-5}.

3. A solution contains 0.10 molal each of Ba^{2+} and Sr^{2+}. The solubility products of $BaCO_3$ and $SrCO_3$ are 2.0×10^{-9} and 5.2×10^{-10}, respectively. Describe what happens as each of the solids is added to the solution.

4. Magnesium hydroxide is slightly soluble in water. If the pH of a saturated solution of $Mg(OH)_2$ is 10.38, find the solubility product of $Mg(OH)_2$.

5. Find the concentration of Ca^{2+} and CO_3^{2-} in air-saturated water (the partial pressure of CO_2 is 3×10^{-4} atm) assuming $K_{sp} = 3.84 \times 10^{-9}$ for $CaCO_3$ and the equilibrium constant $K = 1.4 \times 10^{-6}$ for the reaction

$$CO_2(g) + H_2O(l) + CaCO_3(s) = Ca^{2+} + 2HCO_3^-(aq)$$

Appendices

Appendix I   Heats of formation, standard entropies and heat capacities
Appendix II   Standard free energies of formation
Appendix III Properties of selected elements
Appendix IV   Standard half-cell potentials in aqueous solutions

APPENDIX I

Heats of formation, standard entropies and heat capacities

Substance	Phase	$\Delta H^o_{f,298}$ (kJ mol^{-1})	S^o_{298} (J mol^{-1}K^{-1})	$C_P = a + bT + cT^{-2}$ (J mol^{-1}K^{-1})		
				a	b × 10^3	c × 10^{-5}
Ag	Solid	0	42.70	21.30	8.54	1.51
Ag	Liquid	8.94	47.19	30.54		
AgCl	Solid	-127.1	96.28	62.26	4.18	-11.30
AgBr	Solid	-100.7	107.2	33.19	64.46	
AgI	Solid, α	-61.95	115.5	24.36	100.9	
Ag$_2$O	Solid	-30.56	121.8	59.36	40.81	-4.19
Ag$_2$S	Solid, α	-31.31	143.6	42.40	110.5	
Al	Solid	0	28.34	20.67	12.39	
Al	Liquid	8.23	34.74	31.80		
AlF$_3$	Solid, α	-1511.1	66.52	72.29	45.88	-9.63
AlCl	Gas	-51.49	227.9	37.67		-2.85
AlCl$_3$	Solid	-706.0	109.3	55.46	117.2	
AlCl$_3$	Gas	-584.8	314.5	82.88		-11.05
Al$_2$O$_3$	Solid	-1678.2	51.07	106.7	17.79	-28.55
Al$_2$S$_3$	Solid	-723.8	123.5	102.2	36.08	
AlN	Solid	-318.6	20.18	34.40	16.95	-8.37
As	Solid	0	35.71	23.18	5.52	
As$_4$	Gas	153.4	327.5	82.94	0.13	-5.13
As$_2$O$_3$	Solid	-653.6	122.8	59.83	175.7	
As$_2$S$_3$	Solid	-167.4	163.7	105.7	36.46	
Au	Solid	0	47.39	23.69	5.19	
B	Solid	0	2.99	19.82	5.78	-9.21
B$_2$O$_3$	Solid	-1272.5	54.0	57.06	73.05	-14.06
B$_2$O$_3$	Liquid	-1253.7	78.49	127.7		
BN	Solid	-252.4	14.32	33.91	14.73	-23.06
B$_4$C	Solid	-71.58	27.13	96.24	22.60	-44.87
Ba	Solid	0	62.46	-44.43	158.4	22.49
BaCl$_2$	Solid, α	-1207.7	96.40	92.93	3.18	-16.74
BaO	Solid	-553.8	70.45	53.33	4.35	-8.31
Be	Solid	0	9.50	21.22	5.69	-5.88
BeO	Solid	-608.6	14.15	41.61	10.21	-17.37
Bi	Solid	0	56.72	18.80	22.60	
Bi$_2$O$_3$	Solid, α	-571.0	151.5	103.6	33.49	
Bi$_2$S$_3$	Solid	-201.8	200.5	110.0	41.02	
Br	Gas	111.9	174.9	19.88	1.49	0.42
Br$_2$	Gas	30.9	245.4	37.36	0.46	-1.30
C(graphite)	Solid	0	5.74	17.15	4.27	-8.79

Substance	Phase	$\Delta H^\circ_{f,298}$ (kJ mol^{-1})	S°_{298} (J mol^{-1}K^{-1})	$C_P = a + bT + cT^{-2}$ (J mol^{-1}K^{-1})		
				a	b × 10^3	c × 10^{-5}
CH$_4$	Gas	-74.85	186.2	23.64	47.87	-1.93
C$_2$H$_2$	Gas	226.8	200.9	43.66	31.67	-7.51
CO	Gas	-110.5	197.6	28.41	4.10	-0.46
CO$_2$	Gas	-393.5	213.7	44.14	9.04	-8.54
COS	Gas	-138.4	231.5	47.41	2.61	-7.66
Ca	Solid, α	0	41.6	25.51	2.61	
Ca	Liquid	10.90	50.65	30.13		
CaF$_2$	Solid, α	-1220.2	68.86	59.86	30.47	1.97
CaCl$_2$	Solid	-796.2	104.7	30.06	12.73	-2.51
CaO	Solid	-635.	38.10	49.62	4.52	-7.00
Ca(OH)$_2$	Solid	-985.8	83.43	105.4	11.95	-18.98
CaSO$_4$	Solid	-1434.8	106.7	70.24	98.79	
CaC$_2$	Solid, α	-59.02	70.32	68.65	11.89	-8.67
CaCO$_3$	Solid	-1207.1	88.70	104.5	21.92	-25.94
CaS	Solid	-476.1	56.48	45.19	7.74	
CaSiO$_3$	Solid	-1634.9	82.01	108.2	16.49	-23.64
Ca$_2$SiO$_4$	Solid	-136.9	120.6	151.7	36.96	-30.31
Cd	Solid	0	51.80	22.22	12.30	
Cd	Liquid	5.81	61.05	29.71		
Cd	Gas	111.8	167.6	20.79		
CdCl$_2$	Solid	-391.0	115.3	47.30	91.67	
CdO	Solid	-259.4	54.80	48.24	6.38	-4.90
CdS	Solid	-141.4	69.04	44.56	13.81	
Ce	Solid, α	0	72.00	23.50	10.40	
Ce$_2$O$_3$	Solid	-1822.6	150.7	107.9	41.44	-9.21
Cl$_2$	Gas	0	223.0	36.90	0.25	-2.84
Co	Solid, α	0	30.04	21.39	14.31	-0.88
Co	Solid, β	1.29	32.66	13.81	24.52	
CoO	Solid	-238.9	52.93	48.28	8.54	1.67
CoS	Solid	-121.7	63.95	44.37	10.51	
Cr	Solid	0	23.64	24.44	9.87	-3.68
Cr	Liquid	26.10	36.23	39.33		
Cr$_2$O$_3$	Solid	-1130.2	81.21	119.4	9.21	-15.66
Cr$_3$C$_2$	Solid	-109.7	85.39	125.7	23.36	-31.23
Cr$_7$C$_3$	Solid	-228.1	200.9	238.4	60.86	-42.36
Cs	Liquid	0	85.27	31.90		
Cu	Solid	0	33.15	22.64	6.28	
Cu	Liquid	9.31	36.25	31.38		
Cu	Gas	336.8	166.3	9.93	5.07	
Cu$_2$O	Solid	-167.4	93.09	62.34	23.85	
CuO	Solid	-155.2	42.68	38.79	20.08	

Substance	Phase	$\Delta H^\circ_{f,298}$ (kJ mol^{-1})	S°_{298} (J mol^{-1}K^{-1})	$C_P = a + bT + cT^{-2}$ (J mol^{-1}K^{-1})		
				a	b × 10^3	c × 10^{-5}
Cu$_2$S	Solid, α	-79.50	120.9	81.59		
Cu$_2$S	Solid, β	-79.53	121.0	97.28		
Cu$_2$S	Solid, γ	-79.53	121.0	85.02		
CuS	Solid	-52.30	66.53	44.35	11.05	
CuSO$_4$	Solid	-771.1	109.2	78.53	71.97	
Fe	Solid, α	0	27.28	17.49	24.77	
Fe	Solid, γ	6.78	33.66	26.61	6.28	
Fe	Solid, δ	3.89	29.71	28.28	7.53	
Fe	Liquid	13.13	34.29	35.40	3.75	
Fe	Gas	416.3	180.4	15.72	3.47	
FeCl$_2$	Solid	-342.4	118.1	79.28	8.71	-4.90
Fe$_{0.947}$O	Solid	-264.6	58.81	48.81	8.37	-2.80
FeO	Solid	-264.4	58.79	51.80	6.78	-1.59
Fe$_3$O$_4$	Solid, α	-1116.7	155.5	91.55	201.7	
Fe$_2$O$_3$	Solid, α	-821.3	87.45	98.28	77.82	-14.85
FeS	Solid, α	-100.4	60.29	21.72	110.5	
Fe$_3$C	Solid, α	25.12	104.7	82.21	83.72	
Ga	Solid	0	41.02	26.10		
GaAs	Solid	-81.63	64.26	45.21	6.07	
GaN	Solid	-109.7	29.72	38.09	9.00	
Ga$_2$O$_3$	Solid	-1083.3	84.68	112.9	15.45	-21.01
GaP	Solid	-122.2	52.33	41.86	6.82	0
GaSb	Solid	-41.86	77.36	45.63	12.56	0
Ge	Solid	0	31.10	21.60	5.86	0
GeO$_2$	Solid	-580.2	39.77	66.64	11.60	-17.75
H$_2$	Gas	0	130.6	27.28	3.26	0.50
HBr	Gas	-36.38	198.6	26.15	5.86	1.09
HCl	Gas	-92.31	186.8	26.53	4.60	1.09
HI	Gas	26.36	206.5	26.32	5.94	0.92
H$_2$O	Liquid	-285.8	69.95	75.44		
H$_2$O	Gas	-241.8	188.7	30.00	10.71	0.34
H$_2$S	Gas	-20.50	205.6	32.68	12.39	-1.93
Hf	Solid	0	43.58	23.47	7.62	
HfO$_2$	Solid	-1113.5	59.44	72.79	8.71	-14.57
Hg	Liquid	0	75.93	30.39	-11.47	
HgCl$_2$	Solid	-228.4	140.1	63.96	43.53	
HgO	S,red	90.84	70.32	37.67	25.12	
HgS	S,red	53.37	82.46	43.79	15.57	
I$_2$	Gas	62.43	260.6	37.40	0.57	-0.63
In	Solid	0	57.85	24.32	10.47	
In$_2$O$_3$	Solid	-926.4	108.0	123.9	7.95	-23.06

Substance	Phase	$\Delta H^\circ_{f,298}$ (kJ mol^{-1})	S°_{298} (J mol^{-1}K^{-1})	$C_P = a + bT + cT^{-2}$ (J mol^{-1}K^{-1})		
				A	b × 10^3	c × 10^{-5}
K	Solid	0	64.72	25.28	13.06	
KCl	Solid	-436.9	82.59	41.40	21.77	3.22
K$_2$O	Solid	-363.4	94.19	95.69	-4.94	11.05
K$_2$CO$_3$	Solid	-1150.2	155.5	80.29	109.0	
La	Solid	0	56.93	25.83	6.70	
La$_2$O$_3$	Solid	-1794.1	128.1	120.8	12.89	-13.73
Li	Solid	0	29.09	13.94	34.37	
LiF	Solid	-617.2	35.58	38.26	21.73	
Li$_2$O	Solid	-596.9	37.93	62.54	25.45	-14.15
Mg	Solid	0	42.51	22.30	10.25	-0.43
Mg	Liquid	9.29	148.6	22.05	10.90	
Mg	Gas	147.6	26.95	20.79		
MgF$_2$	Solid	-1123.9	57.26	70.87	10.55	-9.21
MgCl$_2$	Solid	-641.7	89.66	79.12	5.94	-8.62
MgO	Solid	-601.5	26.95	49.00	3.14	-11.72
MgS	Solid	-351.6	50.36	43.12	8.25	
MgCO$_3$	Solid	-1111.7	65.86	77.91	57.74	-17.41
Mn	Solid, α	0	32.02	23.86	14.15	-1.57
MnCl$_2$	Solid	-482.2	118.3	75.52	13.23	-5.73
MnO	Solid	-385.1	59.86	46.51	8.12	-3.68
Mn$_3$O$_4$	Solid, α	-1387.2	154.0	145.0	45.29	-9.21
Mn$_2$O$_3$	Solid	-957.3	110.5	103.5	35.08	-13.56
MnO$_2$	Solid	-520.3	53.16	69.49	10.21	-16.24
MnS	S, green	-213.5	80.37	47.72	7.53	
MnSO$_4$	Solid	-1065.8	112.2	122.5	37.34	-29.47
Mo	Solid	0	28.67	25.57	2.85	-2.18
MoO$_3$	Solid	-1164.1	77.82	84.01	24.70	-15.40
Mo$_2$N	Solid	-69.49	87.91	46.84	57.77	
N$_2$	Gas	0	191.5	27.87	4.27	
NH$_3$	Gas	-45.94	192.7	37.32	18.66	-6.49
N$_2$O	Gas	82.09	220.0	45.71	8.62	-8.54
NO	Gas	90.33	210.8	29.43	3.85	-0.59
Na	Solid	0	51.28	82.51	-369.5	
NaCl	Solid	-412.8	72.17	45.96	16.33	
Na$_2$O	Solid, α	-415.3	75.10	55.51	70.24	-4.14
NaOH	Solid	-426.1	64.46	71.79	-110.9	
Na$_2$SO$_4$	Solid, α	1396.0	149.6	98.37	132.9	
Na$_2$CO$_3$	Solid	-1130.9	138.8	58.49	227.6	-13.10
Na$_3$AlF$_6$	Solid, α	-82.88	238.6	192.4	123.3	-11.64
Nb	Solid	0	36.54	23.72	4.02	
NbO	Solid	-419.9	46.05	42.03	9.84	-3.27

Substance	Phase	$\Delta H^\circ_{f,298}$ (kJ mol^{-1})	S°_{298} (J mol^{-1}K^{-1})	$C_P = a + bT + cT^{-2}$ (J mol^{-1}K^{-1})		
				a	b × 10³	c × 10⁻⁵
NbO$_2$	Solid, α	-795.3	55.67	61.45	25.79	-10.13
Ni	Solid, α	0	29.87	12.54	35.82	2.47
Ni	Solid, β	0.63	30.95	25.10	7.53	
Ni	Liquid	8.32	27.38	38.91		
NiCl$_2$	Solid	-305.6	97.74	73.26	13.23	-4.98
NiO	Solid, α	-240.6	38.08	-20.88	157.2	16.28
NiO	Solid, β	-240.6	38.08	58.10		
NiO	Solid, γ	-240.6	38.08	46.80	8.46	
NiS	Solid, α	-94.14	52.93	43.76	22.20	-2.90
NiSO$_4$	Solid	-873.6	103.9	126.0	41.53	
O$_2$	Gas	0	205.0	29.96	4.18	-1.67
P	S, white	0	41.11	19.13	15.82	
P$_2$O$_5$	Solid	-1492.7	114.5	74.93	162.4	-15.61
Pb	Solid	0	64.80	23.56	9.75	
Pb	Liquid	4.29	71.72	32.43	-3.10	
Pb	Gas	195.6	175.3	6.67	8.96	
PbCl$_2$	Solid	-359.6	136.1	67.39	16.74	
PbCl$_2$	Liquid	-174.1	317.3	118.1		
PbO	S, red	-219.3	65.27	38.20	25.52	
PbO	S, yellow	-217.9	67.36	45.09	12.23	
PbO	Liquid	-195.4	85.97	65.00		
PbS	Gas	131.9	251.5	44.62	16.41	
S	S, ortho	0	32.05	14.81	24.06	0.73
S	S, mono	0.34	32.97	68.35	-118.6	
S$_2$	Gas	128.5	228.1	36.49	0.67	-3.77
SO$_2$	Gas	-296.8	248.1	43.43	10.63	-5.94
SO$_3$	Gas	-395.7	256.7	57.32	26.86	-13.05
Sb	Solid	0	45.54	23.06	7.28	
SbCl$_3$	Gas	-313.7	337.4	43.12	239.0	
Sb$_2$O$_3$	Solid	-708.9	141.1	79.95	71.58	
Sb$_2$S$_3$	Solid	-205.1	182.1	101.9	60.57	
Se	Solid	0	42.28	17.90	25.12	
SeO$_2$	Solid	-225.2	66.72	69.61	3.90	-11.05
Si	Solid	0	18.84	23.94	2.47	-4.14
Si	Liquid	455.9	167.9	25.62		
SiC	Solid	-66.98	16.53	50.78	1.97	-49.23
SiCl$_2$	Gas	-167.4	282.0	57.60	0.38	-5.65
SiCl$_4$	Gas	-663.1	331.0	101.5	6.87	-11.51
SiI$_4$	Solid	-199.3	265.6	82.00	87.49	
SiO$_2$	Solid	-190.7	41.46	46.95	34.31	-11.30
Si$_3$N$_4$	Solid	-745.1	113.0	43.91	1.00	-6.03

Substance	Phase	$\Delta H^\circ_{f,298}$ (kJ mol^{-1})	S°_{298} (J mol^{-1}K^{-1})	$C_P = a + bT + cT^{-2}$ (J mol^{-1}K^{-1})		
				a	b × 10^3	c × 10^{-5}
Sm	Solid	0	69.57	52.62		
Sm$_2$O$_3$	S, cubic			128.3	21.26	-16.58
Sn	S, white	0	51.21	21.59	18.16	
Sn	Liquid	6.63	63.97	34.69	-9.21	
SnO$_2$	Solid	-580.7	52.30	73.89	10.04	-21.59
SnS$_2$	Solid	-153.6	87.49	64.92	17.58	
Ta	Solid	0	41.53	27.84	-2.18	-1.88
Ta$_2$O$_5$	Solid	-2047.0	143.2	154.9	27.46	-24.78
Te	Solid	0	49.52	19.17	21.98	
TeO$_2$	Solid	-323.6	74.09	65.22	14.57	-5.02
Ti	Solid, α	0	30.63	22.09	10.04	
Ti	Solid, β	6,59	38.38	19.83	7.95	
TiO	Solid, α	-542.9	34.74	44.25	15.07	-7.79
TiO	Solid, β	-542.9	34.74	49.60	12.56	
Ti$_2$O$_3$	Solid, α	-1521.6	77.27	30.60	224.0	
Ti$_2$O$_3$	Solid, β	-1521.6	77.27	145.2	5.44	-42.70
Ti$_3$O$_5$	Solid, α	-2460.5	129.5	148.5	123.5	
Ti$_3$O$_5$	Solid, β	-2460.5	129.5	174.1	33.49	
TiO$_2$	S, rutile	-944.8	50.33	75.19	1.17	-18.20
Tl	Solid, α	0	64.21	15.66	25.28	2.80
Tl$_2$O$_3$	Solid	-390.6	137.3	131.9	3.56	-22.27
U	Solid, α	0	50.20	10.92	37.45	4.90
UO$_2$	Solid	-1085.0	77.03	80.33	6.78	-16.57
V	Solid	0	28.95	20.50	10.80	0.84
VN	Solid	-217.3	37.30	45.79	8.79	-9.25
VO	Solid	-432.0	39.01	47.39	13.48	-5.27
VO$_2$	Solid, α	-713.7	51.78	62.62		
VO$_2$	Solid, β	-713.7	51.78	74.72	7.12	-16.53
V$_2$O$_3$	Solid	-1219.4	98.12	122.9	19.93	-22.69
V$_2$O$_5$	Solid	-1551.3	130.6	194.8	-16.33	-55.34
W	Solid	0	32.64	23.81	3.26	
WC	Solid	-38.09	41.86	43.41	8.62	-9.33
WO$_3$	Solid	-843.3	75.93	73.17	28.42	
Y	Solid, α	0	44.50	23.94	7.56	0.33
YN	Solid	-299.3	37.67	45.63	6.49	-7.33
Zn	Solid	0	41.63	22.38	10.04	
ZnCl$_2$	Solid, α	-415.3	111.5	60.70	23.02	
ZnO	Solid	-350.6	43.64	49.00	5.11	-9.12
ZnS	Solid	-205.2	57.66	50.88	5.19	-5.69
ZnSO$_4$	Solid	-981.8	110.6	91.67	76.19	

Substance	Phase	$\Delta H^\circ_{f,298}$ (kJ mol^{-1})	S°_{298} (J mol^{-1}K^{-1})	$C_P = a + bT + cT^{-2}$ (J mol^{-1}K^{-1})		
				A	b × 10^3	c × 10^{-5}
Zr	Solid, α	0	39.00	21.97	11.63	
ZrN	Solid	-368.4	38.89	46.46	7.03	-7.20
ZrO$_2$	Solid, α	-1100.8	50.71	69.62	7.53	-14.06

Source: Data are largely from *Metallurgical Thermochemistry*, O. Kubaschewski and C.B. Alcock, 5th ed. Pergamon Press, 1979.

APPENDIX II

Standard free energies of formation

$$\Delta G^o = \Delta H^o - T\Delta S^o$$

Reaction	ΔH^o (kJ/mol)	ΔS^o (J/mol K)	Temperature Range (K)	Ref
$Ag(s) + \frac{1}{2}Br_2(g) = AgBr(l)$	-97.3	-27.7	715~838	11
$Ag(s) + \frac{1}{2}Cl_2(g) = AgCl(l)$	-106.0	-25.4	803~1193	11
$Ag(s) + \frac{1}{2}F_2(g) = AgF(s)$	-200.8	54.8	298~708(M)	4,7
$Ag(s) + \frac{1}{2}I_2(g) = AgI(l)$	-74.1	-24.1	873~973	11
$2Ag(s) + \frac{1}{2}O_2(g) = Ag_2O(s)$	-28.1	-60.6	298~1000	1
$2Ag(s) + \frac{1}{2}S_2(g) = Ag_2S(s)$	-161.3	168.6	298~1103(M)	4
$Al(l) + \frac{1}{2}Cl_2(g) = AlCl(g)$	-77.4	58.2	933(m)~2273	4
$Al(l) + 1\frac{1}{2}Cl_2(g) = AlCl_3(g)$	-602.1	-67.9	933(m)~2273	4
$Al(l) + 1\frac{1}{2}F_2(g) = AlF_3(s)$	-1,507.7	-257.9	933(m)~1549(s)	4
$Al(l) + 1\frac{1}{2}F_2(g) = AlF_3(g)$	-1,227.8	-78.1	933(m)~2000	2
$Al(l) + \frac{1}{2}N_2(g) = AlN(s)$	-327.1	-115.5	933(m)~2273	4
$Al(l) + 3Na(l) + 3F_2(g) = Na_3AlF_6(l)$	-3,378.2	-623.4	1285(M)~2273	4
$2Al(l) + \frac{1}{2}O_2(g) = Al_2O(g)$	-170.7	49.4	933~2273	15
$2Al(s) + 1\frac{1}{2}O_2(g) = Al_2O_3(s)$	-1,675.1	-313.2	298~933(m)	2
$2Al(l) + 1\frac{1}{2}O_2(g) = Al_2O_3(s)$	-1,682.9	-323.2	933~2315(M)	2
$2Al(l) + 1\frac{1}{2}O_2(g) = Al_2O_3(l)$	-1,574.1	-275.0	2315~2767(b)	2
$2Al(g) + 1\frac{1}{2}O_2(g) = Al_2O_3(l)$	-2,106.4	-468.6	2767~3500	2
$Al_2O_3(s) + SiO_2(s) = Al_2O_3 \cdot SiO_2(s)$	-8.8	-3.9	298~1973	4
$3Al_2O_3(s) + 2SiO_2(s) = 3Al_2O_3 \cdot 2SiO_2(s)$	8.6	17.4	293~2023(M)	15
$Al_2O_3(s) + TiO_2(s) = Al_2O_3 \cdot TiO_2(s)$	-25.3	-3.9	298~2133(M)	8
$CaO(s) + Al_2O_3(s) = CaO \cdot Al_2O_3(s)$	-18.0	19.0	773~1878(M)	18
$3CaO(s) + Al_2O_3(s) = 3CaO \cdot Al_2O_3(s)$	-12.6	24.7	773~1808(M)	18
$CaO(s) + 2Al_2O_3(s) = CaO \cdot 2Al_2O_3(s)$	-17.0	25.5	773~2023(M)	18

Reaction	ΔH^o (kJ/mol)	ΔS^o (J/mol K)	Temperature Range (K)	Ref
$3Li(l) + Al(l) + 3F_2(g) = Li_3AlF_6(l)$	-3,240.0	-400	1058(M)~1615(M)	15
$MgO(s) + Al_2O_3(s) = MgO \cdot Al_2O_3(s)$	-35.6	2.1	298~1673	15, 18
$MnO(s) + Al_2O_3(s) = MnO \cdot Al_2O_3(s)$	-48.1	-7.3	773~1473	18
$Na_2O(s) + Al_2O_3(s) = Na_2O \cdot Al_2O_3(s)$	-185.0	2.9	773~1404	18
$CaO(s) + Al_2O_3(s) + SiO_2(s) = CaO \cdot Al_2O_3 \cdot SiO_2(s)$	-105.9	-14.2	298~1673	4
$CaO(s) + Al_2O_3(s) + 2SiO_2(s) = CaO \cdot Al_2O_3 \cdot 2SiO_2(s)$	-139.0	-17.2	298~1826	8
$2CaO(s) + Al_2O_3(s) + SiO_2(s) = 2CaO \cdot Al_2O_3 \cdot SiO_2(s)$	-170.0	-8.8	298~1773	8
$3CaO(s) + Al_2O_3(s) + 3SiO_2(s) = 3CaO \cdot Al_2O_3 \cdot 3SiO_2(s)$	-389.0	-100.0	298~1673	8
$Ga(l) + ¼As_4(g) = GaAs(s)$	-115.0	-72.0	303~1238	12
$In(l) + ¼As_4(g) = InAs(s)$	-99.3	-64.8	430~1215	12
$4B(s) + C(s) = B_4C(s)$	-41.5	-5.6	298~2303	15
$B(s) + ½N_2(g) = BN(s)$	-250.6	-87.6	298~2453(m)	4
$B(s) + ½O_2(g) = BO(g)$	-3.8	88.8	298~2303	15
$2B(s) + 1½O_2(g) = B_2O_3(g)$	-1,229.0	-210	723~2316(B)	15
$Ba(l) + C(s) + 1½O_2(g) = BaCO_3(s)$	-1,203.3	-249.2	1073~1333	4
$Ba(l) + Cl_2(g) = BaCl_2(l)$	-811.3	-121.5	1235(M)~1235(M)	15
$Ba(l) + F_2(g) = BaF_2(l)$	-1,154.0	-129	1641~1895	15
$Ba(s) + H_2(g) + O_2(g) = Ba(OH)_2(s)$	-941.3	-291.1	298~681(M)	7,8
$Ba(l) + H_2(g) + O_2(g) = Ba(OH)_2(l)$	-918.4	-248.4	1002(m)~1263	7,8
$Ba(s) + ½O_2(g) = BaO(s)$	-568.2	-97.0	298~1002(m)	3
$Ba(l) + ½O_2(g) = BaO(s)$	-557.2	-102.7	1002~1895(b)	4
$Ba(l) + ½S_2(g) = BaS(s)$	-543.9	-123.4	1002(m)~1895(b)	4
$3Be(l) + N_2(g) = Be_3N_2(s)$	-616.0	-203.0	1560~2473(M)	15
$Be(s) + ½O_2(g) = BeO(s)$	-608.0	-97.7	298~1560	15
$Be(l) + ½O_2(g) = BeO(s)$	-613.6	-100.9	1560~2273	15
$Be(s) + ½S_2(g) = BeS(s)$	-300.0	-86.6	298~1560	8

Reaction	ΔH^o (kJ/mol)	ΔS^o (J/mol K)	Temperature Range (K)	Ref
$2Bi(l) + 1\frac{1}{2}O_2(g) = Bi_2O_3(s)$	-590.2	-292.6	545~1097(M)	1
$2Bi(l) + 1\frac{1}{2}O_2(g) = Bi_2O_3(l)$	-445.2	-159.6	1097(M)~1773	4
$2Bi(l) + 1\frac{1}{2}S_2(g) = Bi_2S_3(s)$	-360.0	-274.0	545~1050(M)	8
$C(s) + 2Cl_2(g) = CCl_4(g)$	-89.1	-129.2	298~2273	15
$C(s) + 2F_2(g) = CF_4(g)$	-933.2	-151.5	298~2273	15
$C(s) + 2H_2(g) = CH_4(g)$	-87.4	-108.7	298~2500	2
$C(s) + \frac{1}{2}N_2(g) = CN(g)$	433.5	99.6	298~2273	15
$C(s) + \frac{1}{2}O_2(g) = CO(g)$	-112.9	86.5	298~2500	2
$C(s) + O_2(g) = CO_2(g)$	-394.8	0.836	298~2500	2
$C(s) + \frac{1}{2}S_2(g) = CS(g)$	163.2	87.9	298~2273	4
$C(s) + S_2(g) = CS_2(g)$	-11.4	6.5	298~2273	4
$Ca(l) + 2C(s) = CaC_2(s)$	-60.3	26.3	1115~1755	15
$Ca(s) + C(s) + 1\frac{1}{2}O_2(g) = CaCO_3(s)$	-1,196.3	-242.1	298~1112(m)	3
$Ca(l) + C(s) + 1\frac{1}{2}O_2(g) = CaCO_3(s)$	-1,196.2	-245.0	1112~1473	4
$Ca(s) + Cl_2(g) = CaCl_2(s)$	-794.5	-142.3	298~1045(M)	3
$Ca(l) + Cl_2(g) = CaCl_2(l)$	-798.6	-146.0	1112(m)~1764(b)	4
$Ca(s) + F_2(g) = CaF_2(s)$	-1,221.8	-164.9	298~1112(m)	2
$Ca(l) + F_2(g) = CaF_2(s)$	-1,212.6	-156.7	1112(m)~1691(M)	2
$Ca(s) + H_2(g) + O_2(g) = Ca(OH)_2(s)$	-983.1	-285.2	298~1000	7,8
$Ca(s) + \frac{1}{2}O_2(g) = CaO(s)$	-633.1	-99.0	298~1112(m)	3
$Ca(l) + \frac{1}{2}O_2(g) = CaO(s)$	-640.2	-108.6	1112~1764(b)	4
$Ca(g) + \frac{1}{2}O_2(g) = CaO(s)$	-795.4	-195.1	1764~2500	3
$3Ca(s) + P_2(g) = Ca_3P_2(s)$	-650.0	-216.0	298~1115	8
$Ca(s,\alpha) + \frac{1}{2}S_2(g) = CaS(s)$	-541.6	-95.4	298~721	3
$Ca(s,\beta) + \frac{1}{2}S_2(g) = CaS(s)$	-542.6	-96.1	721~1112(m)	3
$Ca(l) + \frac{1}{2}S_2(g) = CaS(s)$	-548.1	-103.8	1112~1764(b)	4
$CaO(s) + SO_2(g) + \frac{1}{2}O_2(g) = CaSO_4(s,\alpha)$	-454.0	-232.0	1468~1638(M)	16
$CaO(s) + SO_2(g) + \frac{1}{2}O_2(g) = CaSO_4(s,\beta)$	-460.0	-238.0	1223~1468	16
$CaO(s) + Al_2O_3(s) = CaO \cdot Al_2O_3(s)$	-18.0	19.0	773~1878(M)	18

Reaction	ΔH^o (kJ/mol)	ΔS^o (J/mol K)	Temperature Range (K)	Ref
$CaO(s) + 2Al_2O_3(s) = CaO \cdot 2Al_2O_3(s)$	-17.0	25.5	773~2023(M)	18
$3CaO(s) + Al_2O_3(s) = 3CaO \cdot Al_2O_3(s)$	-12.6	24.7	773~1808(M)	18
$CaO(s) + Fe_2O_3(s) = CaO \cdot Fe_2O_3(s)$	-30.0	4.8	973~1489(M)	18
$2CaO(s) + Fe_2O_3(s) = 2CaO \cdot Fe_2O_3(s)$	-53.1	2.5	973~1723(M)	18
$2CaO(s) + P_2(g) + 2½O_2(g) = 2CaO \cdot P_2O_5(s)$	-2,190.0	-586.0	298~1626(M)	8
$2CaO(s) + SiO_2(s) = 2CaO \cdot SiO_2(s)$	-120.0	11.3	298~2403(M)	8
$3CaO(s) + 2SiO_2(s) = 3CaO \cdot 2SiO_2(s)$	-237.0	-9.6	298~1773	8
$CaO(s) + TiO_2(s) = CaO \cdot TiO_2(s)$	-80.0	-3.4	298~1673	18
$3CaO(s) + 2TiO_2(s) = 3CaO \cdot 2TiO_2(s)$	-270.0	11.5	298~1673	18
$4CaO(s) + 3TiO_2(s) = 4CaO \cdot 3TiO_2(s)$	-293.0	17.6	298~1673	18
$3CaO(s) + P_2(g) + 2½O_2(g) = 3CaO \cdot P_2O_5(s)$	-2,314.0	-556.0	298~2003(M)	8
$3CaO(s) + SiO_2(s) = 3CaO \cdot SiO_2(s)$	-118.8	6.7	298~1773	4
$2CaO(s) + SiO_2(s) = 2CaO \cdot SiO_2(s)$	-118.8	11.3	298~2403(M)	4
$CaO(s) + SiO_2(s) = CaO \cdot SiO_2(s)$	-92.8	-2.5	298~1813(M)	4
$3CaO(s) + 2SiO_2(s) = 3CaO \cdot 2SiO_2(s)$	-236.8	-9.6	298~1773	4
$Cd(l) + Cl_2(g) = CdCl_2(s)$	-389.6	-153.0	594(m)~8421(M)	4
$Cd(l) + Cl_2(g) = CdCl_2(l)$	-352.6	-110.3	841(M)~1040(b)	14
$Cd(l) + ½O_2(g) = CdO(s)$	-263.2	-104.9	594(m)~1040(b)	4
$Cd(g) + ½O_2(g) = CdO(s)$	-356.7	-198.5	1040~1500	5
$Ce(s) + O_2(g) = CeO_2(s)$	-1,084.0	-212.0	298~1071	20
$2Ce(s) + 1½O_2(g) = Ce_2O_3(s)$	-1,788.0	-286.6	298~1071(m)	4
$Ce(l) + ½S_2(g) = CeS(s)$	-534.9	-91.0	1071(m)~2723(M)	4
$Co(s) + ½O_2(g) = CoO(s)$	-233.9	-70.7	298~1400	3
$Co(s) + S_2(g) = CoS_2(s)$	-280.3	182.4	298~872	4

Reaction	ΔH^o (kJ/mol)	ΔS^o (J/mol K)	Temperature Range (K)	Ref
$3Cr(s) + 2C(s) = Cr_3C_2(s)$	-79.1	17.7	298~2130(m)	4
$7Cr(s) + 3C(s) = Cr_7C_3(s)$	-153.6	37.2	298~2130(m)	4
$23Cr(s) + 6C(s) = Cr_{23}C_6(s)$	-309.6	77.4	298~1773	4
$Cr(s) + \frac{1}{2}O_2(g) = CrO(l)$	-334.2	63.8	1938~2023(M)	4
$2Cr(s) + 1\frac{1}{2}O_2(g) = Cr_2O_3(s)$	-1,110.1	-249.3	1173~1923	4
$3Cr(s) + 2O_2(g) = Cr_3O_4(s)$	-1,355.0	-265.0	1923~1938(M)	21
$Cu(s) + \frac{1}{2}O_2(g) = CuO(s)$	-152.5	-85.3	298~1356(m)	2
$2Cu(s) + \frac{1}{2}O_2(g) = Cu_2O(s)$	-168.4	-71.2	298~1356(m)	4
$2Cu(l) + \frac{1}{2}O_2(g) = Cu_2O(s)$	-181.7	-80.6	1356~1509(M)	2
$2Cu(l) + \frac{1}{2}O_2(g) = Cu_2O(l)$	-118.7	-39.5	1509~2273	2
$Cu(s) + \frac{1}{2}S_2(g) = CuS(s)$	-115.6	-76.1	298~708	4
$2Cu(s) + \frac{1}{2}S_2(g) = Cu_2S(s, \gamma)$	-140.7	-43.3	298~708	4
$2Cu(s) + \frac{1}{2}S_2(g) = Cu_2S(s, \alpha)$	-131.8	-30.8	708~1356(m)	4
$Cu(s) + Fe(s) + S_2(g) = CuFeS_2(s,a)$	-278.6	-115.3	830~973	4
$3Fe(s, \alpha) + C(s) = Fe_3C(s)$	29.0	28.0	298~1000	4
$3Fe(s, \gamma) + C(s) = Fe_3C(s)$	11.2	11.0	1000~1410	4
$Fe(s, \alpha) + Cl_2(g) = FeCl_2(s)$	-339.4	-119.2	298~950(M)	3
$Fe(s) + Cl_2(g) = FeCl_2(l)$	-286.4	-63.7	950(M)~1297(B)	3
$Fe(s) + Cl_2(g) = FeCl_2(g)$	-169.6	26.5	1297(B)~1809(m)	3
$Fe(s) + 1\frac{1}{2}Cl_2(g) = FeCl_3(s)$	-396.5	-210.4	298~577(M)	2
$Fe(s) + 1\frac{1}{2}Cl_2(g) = FeCl_3(g)$	-261.3	-28.0	605(B)~1809(m)	2
$0.947Fe(s) + \frac{1}{2}O_2(g) = Fe_{0.947}O(s)$	-263.7	-64.4	298~1643(M)	15
$Fe(s) + \frac{1}{2}O_2(g) = \text{"FeO"}(s)$	-264.0	-64.6	298~1650	2
$Fe(l) + \frac{1}{2}O_2(g) = FeO(l)$	-256.0	-53.7	1644(M)~2273	4
$2Fe(s) + 1\frac{1}{2}O_2(g) = Fe_2O_3(s)$	-815.0	-251.1	298~1735	2
$3Fe(s) + 2O_2(g) = Fe_3O_4(s)$	-1,103.1	-307.4	298~1870(M)	2
$Fe(s, \gamma) + \frac{1}{2}S_2(g) = FeS(s)$	-154.9	-56.9	1179~1261	4
$Fe(s) + \frac{1}{2}S_2(g) = FeS(l)$	-164.0	-61.1	1261~1468(M)	4
$Fe(s) + S_2(g) = FeS_2(s)$	-336.9	-244.5	903~1033	4
$Cu(s) + Fe(s) + S_2(g) = CuFeS_2(s, \alpha)$	-278.6	-115.3	830~973	4

Reaction	ΔH^o (kJ/mol)	ΔS^o (J/mol K)	Temperature Range (K)	Ref
$Ga(l) + \frac{1}{4}As_4(g) = GaAs(s)$	-115.0	-72.0	303~1238	12
$Ga(l) + \frac{1}{2}Cl_2(g) = GaCl(g)$	-79.6	52.4	303~2000	12
$Ga(l) + 1\frac{1}{2}Cl_2(g) = GaCl_3(g)$	-442.0	84.7	351(M)~575(B)	3,8
$2Ga(l) + 1\frac{1}{2}O_2(g) = Ga_2O_3(s)$	-1,089.9	-323.6	303(m)~2068(M)	4
$Ga(l) + \frac{1}{4}P_4(g) = GaP(s)$	-142.3	-77.0	303~1790(M)	7,8
$Ga(l) + \frac{1}{2}S_2(g) = GaS(s)$	-276.0	-111.0	303~1233(M)	8
$2Ga(l) + 1\frac{1}{2}S_2(g) = Ga_2S_3(s)$	-719.6	-318.4	303(m)~1363(M)	4
$Ge(s) + O_2(g) = GeO_2(s)$	-575.0	-188.0	298~1210	8
$\frac{1}{2}H_2(g) + \frac{1}{2}Br_2(g) = HBr(g)$	-53.6	6.9	298~2273	4
$\frac{1}{2}H_2(g) + \frac{1}{2}Cl_2(g) = HCl(g)$	-94.1	6.4	298~2273	4
$\frac{1}{2}H_2(g) + \frac{1}{2}F_2(g) = HF(g)$	-274.5	3.5	298~2273	4
$\frac{1}{2}H_2(g) + \frac{1}{2}I_2(g) = HI(g)$	-4.2	8.8	298~2273	4
$H_2(g) + \frac{1}{2}O_2(g) = H_2O(g)$	-247.4	-55.8	298~2500	2
$\frac{1}{2}H_2(g) + \frac{1}{2}S_2(g) = HS(g)$	79.7	15.5	718~2273	2
$H_2(g) + \frac{1}{2}S_2(g) = H_2S(g)$	-91.6	-50.6	298~2273	4
$Hf(s) + O_2(g) = HfO_2(s,a)$	-1,060.0	-174.0	298~1973	7
$Hg(l) + \frac{1}{2}O_2(g) = HgO(s,red)$	-90.8	-70.3	298~773	15
$Hg(l) + \frac{1}{2}S_2(g) = HgS(s,red)$	-53.0	-82.0	298~618	3
$HgS(s,black) = HgS(s,red)$	-4.0	-6.3	618~618	3
$In(l) + \frac{1}{4}As_4(g) = InAs(s)$	-99.3	-64.8	430~1215	12
$In(l) + \frac{1}{2}Cl_2(g) = InCl(l)$	-169.1	-37.1	498(M)~881(B)	7,8
$In(l) + \frac{1}{2}Cl_2(g) = InCl(g)$	-87.2	56.3	430(m)~2000	12
$In(l) + 1\frac{1}{2}Cl_2(g) = InCl_3(s)$	-533.4	-242.4	430(m)~856(M)	7,8
$In(l) + 1\frac{1}{2}Cl_2(g) = InCl_3(g)$	-375.0	-36.7	500~800	3,8
$In(l) + \frac{1}{4}P_4(g) = InP(s)$	-92.4	-74.1	430~1328(M)	7,8
$Ir(s) + O_2(g) = IrO_2(s)$	-234.0	-170.0	298~1273	7
$Ir(s) + S_2(g) = IrS_2(s)$	-268.0	-190.0	298~1273	8
$2K(l) + C(s) + 1\frac{1}{2}O_2(g) = K_2CO_3(s)$	-1,149.3	-288.5	336(m)~1037(b)	2
$2K(g) + C(s) + 1\frac{1}{2}O_2(g) = K_2CO_3(s)$	-1,277.1	-410.4	1037~1174(M)	2

Reaction	ΔH^o (kJ/mol)	ΔS^o (J/mol K)	Temperature Range (K)	Ref
$2K(g) + C(s) + 1\frac{1}{2}O_2(g) = K_2CO_3(l)$	-1,204.8	-353.7	1174~2500	2
$K(l) + \frac{1}{2}Cl_2(g) = KCl(s)$	-438.9	-100.4	336(m)~1037(b)	3
$K(g) + \frac{1}{2}Cl_2(g) = KCl(l)$	-474.0	-131.8	1044(M)~1710(B)	4
$K(g) + \frac{1}{2}Cl_2(g) = KCl(g)$	-306.3	-35.5	1710(B)~2273	4
$K(l) + \frac{1}{2}H_2(g) + \frac{1}{2}O_2(g) = KOH(l)$	-402.3	-118.0	673(M)~1037(b)	2
$K(g) + \frac{1}{2}H_2(g) + \frac{1}{2}O_2(g) = KOH(l)$	-469.6	-182.9	1037~1600(B)	2
$2K(l) + \frac{1}{2}O_2(g) = K_2O(s)$	-363.2	-140.4	336(m)~1037(b)	2
$2K(g) + \frac{1}{2}O_2(g) = K_2O(s)$	-478.7	-253.0	1037~2000	2
$2K(l) + \frac{1}{2}S_2(g) = K_2S(s)$	-481.2	-143.5	336~1037(b)	4
$2K(g) + \frac{1}{2}S_2(g) = K_2S(s)$	-633.1	-289.8	1037~1221(M)	4
$2K(g) + \frac{1}{2}S_2(g) = K_2S(l)$	-616.9	-276.6	1221~2000	4,8
$2La(s) + 1\frac{1}{2}O_2(g) = La_2O_3(s)$	-1,790.0	-278.0	298~1193	8
$La(s) + \frac{1}{2}S_2(g) = LaS(s)$	-527.0	-104.0	1193~1773	8
$2La(s) + 1\frac{1}{2}S_2(g) = La_2S_3(s)$	-1,420.0	-286.0	1193~1773	8
$Li(l) + \frac{1}{2}F_2(g) = LiF(l)$	-583.4	-66.8	1121(M)~1620(b)	4
$Li(g) + \frac{1}{2}F_2(g) = LiF(g)$	-500.0	-43.0	1990(B)~2273	15
$2Li(l) + \frac{1}{2}O_2(g) = Li_2O(s)$	-603.8	-136.6	454(m)~1620(b)	2
$2Li(g) + \frac{1}{2}O_2(g) = Li_2O(s)$	-854.7	-290.8	1620~1843(M)	2
$2Li(l) + \frac{1}{2}S_2(g) = Li_2S(s)$	-514.6	-121.3	454(m)~1273	4,8
$Mg(l) + Cl_2(g) = MgCl_2(l)$	-596.8	-114.2	987(M)~1378(b)	2
$Mg(g) + Cl_2(g) = MgCl_2(l)$	-649.0	-157.7	987(M)~1710(B)	15
$Mg(s) + Cl_2(g) = MgCl_2(s)$	-637.0	-155.4	298~923	15
$Mg(l) + F_2(g) = MgF_2(s)$	-1,126.8	-177.8	922(m)~1276	3
$Mg(g) + F_2(g) = MgF_2(l)$	-1,172.0	-215.6	1536(M)~2536(B)	15
$Mg(s) + F_2(g) = MgF_2(s)$	-1,120.0	-171.2	298~922(m)	15
$Mg(s) + C(s) + 1\frac{1}{2}O_2(g) = MgCO_3(s)$	-1,109.5	-274.4	298~922(m)	2
$Mg(s) + H_2(g) + O_2(g) = Mg(OH)_2(s)$	-922.9	-300.8	298~922(m)	2
$Mg(s) + \frac{1}{2}O_2(g) = MgO(s)$	-601.2	-107.6	298~922(m)	2
$Mg(l) + \frac{1}{2}O_2(g) = MgO(s)$	-609.6	-116.5	922~1378(b)	2

Reaction	ΔH^o (kJ/mol)	ΔS^o (J/mol K)	Temperature Range (K)	Ref
$Mg(g) + \frac{1}{2}O_2(g) = MgO(s)$	-732.7	-206.0	1378~2000	2
$Mg(s) + \frac{1}{2}S_2(g) = MgS(s)$	-409.6	-94.4	298~922(m)	4,8
$Mg(l) + \frac{1}{2}S_2(g) = MgS(s)$	-408.9	-98.0	922~1378(b)	2,4
$Mg(g) + \frac{1}{2}S_2(g) = MgS(s)$	-539.7	-193.0	1378~1973	4,8
$MgO(s) + Al_2O_3(s) = MgO \cdot Al_2O_3(s)$	-35.6	2.1	298~1673	15
$MgO(s) + SiO_2(s) = MgO \cdot SiO_2(s)$	-41.1	-6.1	298~1850(M)	4
$2MgO(s) + SiO_2(s) = 2MgO \cdot SiO_2(s)$	-67.2	-4.3	298~2171(M)	4
$2MgO(s) + TiO_2(s) = 2MgO \cdot TiO_2(s)$	-26.4	-1.3	298~1773	15
$3Mn(s) + C(s) = Mn_3C(s)$	-13.9	-1.1	298~1310	4,7
$7Mn(s) + 3C(s) = Mn_7C_3(s)$	-127.6	-21.1	298~1473	4,8
$Mn(s) + Cl_2(g) = MnCl_2(s)$	-478.2	-127.7	298~923(M)	4
$Mn(s) + Cl_2(g) = MnCl_2(l)$	-440.6	-86.9	923(M)~1200	4
$Mn(s) + \frac{1}{2}O_2(g) = MnO(s)$	-388.9	-76.3	298~1517(m)	4
$3Mn(s) + 2O_2(g) = Mn_3O_4(s)$	-1,384.9	-344.4	298~1517(m)	4
$2Mn(s) + 1\frac{1}{2}O_2(g) = Mn_2O_3(s)$	-953.9	-255.2	298~1517(m)	4
$Mn(s) + O_2(g) = MnO_2(s)$	-519.0	-181.0	298~783	18
$Mn(s) + \frac{1}{2}S_2(g) = MnS(s)$	-296.0	-76.7	973~1473	4
$Mn(s) + Si(s) = MnSi(s)$	-61.5	-6.3	298~1519	8
$MnO(s) + Al_2O_3(s) = MnO \cdot Al_2O_3(s)$	-48.1	-7.3	773~1473	18
$MnO(s) + SiO_2(s) = MnO \cdot SiO_2(s)$	-28.0	-2.8	298~1564(M)	4
$2MnO(s) + SiO_2(s) = 2MnO \cdot SiO_2(s)$	-53.6	-24.7	298~1618(M)	4
$Mo(s) + C(s) = MoC(s)$	-7.5	5.4	298~973	4,7
$2Mo(s) + C(s) = Mo_2C(s)$	-45.6	4.2	298~1373	4,7
$Mo(s) + 2Cl_2(g) = MoCl_4(s)$	-472.4	-236.4	298~590(M)	2
$2Mo(s) + \frac{1}{2}N_2(g) = Mo_2N(s)$	-60.7	-14.6	298~773	8
$Mo(s) + O_2(g) = MoO_2(s)$	-578.2	-166.5	298~2273	4
$Mo(s) + O_2(g) = MoO_2(g)$	-18.4	33.9	298~2273	4
$Mo(s) + 1\frac{1}{2}O_2(g) = MoO_3(s)$	-740.2	-246.7	298~1074(m)	4

Reaction	ΔH^o (kJ/mol)	ΔS^o (J/mol K)	Temperature Range (K)	Ref
$Mo(s) + 1\tfrac{1}{2}O_2(g) = MoO_3(l)$	-664.5	-176.6	1074~2000	2
$Mo(s) + 1\tfrac{1}{2}O_2(g) = MoO_3(g)$	-359.8	-59.4	298~2273	4
$2Mo(s) + 1\tfrac{1}{2}S_2(g) = Mo_2S_3(s)$	-594.1	-265.3	298~1473	4,8
$Mo(s) + S_2(g) = MoS_2(s)$	-397.5	-182.0	298~1458(M)	4,8
$Na(l) + \tfrac{1}{2}Cl_2(g) = NaCl(s)$	-411.6	-93.1	371(m)~1074(M)	4
$Na(g) + \tfrac{1}{2}Cl_2(g) = NaCl(l)$	-464.4	-133.9	1074(M)~1738(B)	4
$Na(l) + \tfrac{1}{2}F_2(g) = NaF(s)$	-576.6	-105.5	371(m)~1269(M)	4
$Na(g) + \tfrac{1}{2}F_2(g) = NaF(l)$	-624.3	-148.2	1269(M)~2060(B)	4
$3Na(g) + Al(l) + 3F_2(g) = Na_3AlF_6(l)$	-3,378.2	-623.4	1285(M)~2273	4
$2Na(l) + C(s) + 1\tfrac{1}{2}O_2(g) = Na_2CO_3(s)$	-1,127.5	-273.6	371(m)~1156(b)	2
$2Na(g) + C(s) + 1\tfrac{1}{2}O_2(g) = Na_2CO_3(l)$	-1,229.6	-362.5	1123(M)~2500	2
$Na(l) + \tfrac{1}{2}H_2(g) + \tfrac{1}{2}O_2(g) = NaOH(l)$	-408.1	-125.7	592(M)~1156(b)	2
$Na(g) + \tfrac{1}{2}H_2(g) + \tfrac{1}{2}O_2(g) = NaOH(l)$	-486.6	-192.5	1156(b)~1663(B)	2
$2Na(l) + \tfrac{1}{2}O_2(g) = Na_2O(s)$	-421.6	-141.3	371(m)~1156(b)	4
$2Na(g) + \tfrac{1}{2}O_2(g) = Na_2O(s)$	-571.7	-269.8	1156~1405(M)	2
$2Na(g) + \tfrac{1}{2}O_2(g) = Na_2O(l)$	-519.8	-234.3	1405~2223	2
$2Na(l) + \tfrac{1}{2}S_2(g) = Na_2S(s)$	-394.0	-83.7	371(m)~1156(b)	2
$2Na(g) + \tfrac{1}{2}S_2(g) = Na_2S(s)$	-521.2	-200.1	1156~1223(M)	2
$2Na(g) + \tfrac{1}{2}S_2(g) = Na_2S(l)$	-610.9	-274.7	1223~2000	2
$\tfrac{1}{2}N_2(g) + 1\tfrac{1}{2}H_2(g) = NH_3(g)$	-53.7	-116.5	298~2273	2,4
$\tfrac{1}{2}N_2(g) + \tfrac{1}{2}O_2(g) = NO(g)$	90.4	12.7	298~2273	4
$\tfrac{1}{2}N_2(g) + O_2(g) = NO_2(g)$	32.3	-63.3	298~2273	15
$2Nb(s) + C(s) = Nb_2C(s)$	-194.0	-11.7	298~1773	7
$Nb(s) + 0.98C(s) = NbC_{0.98}(s)$	-137.0	-2.4	298~1773	15
$Nb(s) + O_2(g) = NbO_2(s)$	-784.0	-167	298~2175(M)	15
$2Nb(s) + 2\tfrac{1}{2}O_2(g) = Nb_2O_5(s)$	-1,888.0	-420	298~1785(M)	15
$Nb(s) + \tfrac{1}{2}O_2(g) = NbO(s)$	-415.0	-87.0	298~2210	15
$3Ni(s) + C(s) = Ni_3C(s)$	39.7	17.1	298~773	4,8
$Ni(s) + Cl_2(g) = NiCl_2(s)$	-305.4	-146.4	298~1260	4

Reaction	$\Delta H°$ (kJ/mol)	$\Delta S°$ (J/mol K)	Temperature Range (K)	Ref
$Ni(s) + \tfrac{1}{2}S_2(g) = NiS(l)$	-111.7	-43.6	1067~1728(m)	9,13
$Ni(s) + \tfrac{1}{2}O_2(g) = NiO(s)$	-235.8	-86.2	298~1728(m)	6
$3Ni(s) + S_2(g) = Ni_3S_2(s)$	-336.7	-162.9	298~1064(M)	9,13
$3Ni(s) + S_2(g) = Ni_3S_2(l)$	-237.3	-62.4	1064~1728(m)	9,13
$Ni(s) + \tfrac{1}{2}S_2(g) = NiS(s)$	-153.6	-83.6	298~1067(M)	9,13
$NiO(s) + SO_2(g) + \tfrac{1}{2}O_2(g) = NiSO_4(s)$	-347.5	-293.2	873~1133	17
$\tfrac{1}{4}P_4(g) = P(s,red)$	-32.1	-45.6	298~704	15
$4P_{red}(s) = P_4(g)$	128.5	182.6	298~704	4
$\tfrac{1}{2}P_4(g) = P_2(g)$	108.6	69.5	298~1973	4
$\tfrac{1}{2}P_2(g) + 1\tfrac{1}{2}H_2(g) = PH_3(g)$	-71.5	-108.2	298~1973	4,8
$2P_2(g) + 5O_2(g) = P_4O_{10}(g)$	-3,155.0	-1011	631~1973	15
$\tfrac{1}{2}P_2(g) + O_2(g) = PO_2(g)$	-386.0	-60.3	298~1973	15
$\tfrac{1}{2}P_2(g) + \tfrac{1}{2}O_2(g) = PO(g)$	-77.8	11.6	298~1973	15
$Pb(l) + Cl_2(g) = PbCl_2(l)$	-324.6	-103.5	774(M)~1226(B)	4
$Pb(l) + Cl_2(g) = PbCl_2(g)$	-188.3	7.5	1226(B)~2023(b)	4
$Pb(l) + \tfrac{1}{2}O_2(g) = PbO\ (s,red)$	-221.5	-104.6	600(m)~762(t)	2
$Pb(l) + \tfrac{1}{2}O_2(g) = PbO\ (s,yellow)$	-218.1	-100.2	762~1170(M)	2
$Pb(l) + \tfrac{1}{2}O_2(g) = PbO(s)$	-219.1	-101.2	600~1170(M)	4
$Pb(l) + \tfrac{1}{2}O_2(g) = PbO(l)$	-185.1	-72.0	1170~1789(B)	15
$3Pb(l) + 2O_2(g) = Pb_3O_4(s)$	-702.5	-368.9	600~1473	4
$Pb(l) + \tfrac{1}{2}S_2(g) = PbS(s)$	-163.2	-88.0	600(m)~1386(M)	4
$Pb(l) + \tfrac{1}{2}S_2(g) = PbS(g)$	59.0	54.0	1100~1400	5
$PbO(s) + SO_2(g) + \tfrac{1}{2}O_2(g) = PbSO_4(s)$	-401.0	-262.0	298~1363(M)	7
$S(l) = S(s)$	-1.7	-4.4	388(M)~388(M)	15
$\tfrac{1}{2}S_2(g) = S(l)$	-58.6	-68.3	388(M)~718(B)	15
$\tfrac{1}{2}S_2(g) = S(g)$	217.0	59.6	298~1973	2
$2S(l) = S_2(g)$	120.0	139.6	388(m)~718(b)	10
$2S(g) = S_2(g)$	-469.3	-161.3	298~1973	
$1\tfrac{1}{2}S_2(g) = S_3(g)$	-56.3	-80.4	298~800	10
$2S_2(g) = S_4(g)$	-117.9	-154.7	298~800	10

Reaction	ΔH^o (kJ/mol)	ΔS^o (J/mol K)	Temperature Range (K)	Ref
$2S_2(g) = S_4(g)$	-62.8	-115.5	298~1973	4
$2\frac{1}{2}S_2(g) = S_5(g)$	-203.0*	-240.0	298~800	10
$3S_2(g) = S_6(g)$	-276.1	-305.0	298~1973	4
$3\frac{1}{2}S_2(g) = S_7(g)$	-331.6	-374.1	298~800	10
$4S_2(g) = S_8(g)$	-397.5	-448.1	298~1973	4
$\frac{1}{2}S_2(g) + \frac{1}{2}O_2(g) = SO(g)$	-57.8	5.0	718(b)~2273	4
$\frac{1}{2}S_2(g) + O_2(g) = SO_2(g)$	-361.7	-72.7	718(b)~2273	4
$\frac{1}{2}S_2(g) + 1\frac{1}{2}O_2(g) = SO_3(g)$	-457.9	-163.3	718(b)~2273	4
$2Sb(s) + 1\frac{1}{2}O_2(g) = Sb_2O_3(s)$	-687.6	-241.1	298~904(m)	3
$2Sb(l) + 1\frac{1}{2}O_2(g) = Sb_2O_3(l)$	-660.7	-198.1	929(M)~1860(b)	4
$\frac{1}{2}Se_2(g) + O_2(g) = SeO_2(g)$	-178.0	-66.1	958~1973	8
$\frac{1}{2}Se_2(g) + \frac{1}{2}O_2(g) = SeO(g)$	-9.2	-4.2	958~1973	8
$Si(s) + C(s) = SiC(s,\beta)$	-73.1	-7.7	29 8~1685(m)	4
$Si(l) + C(s) = SiC(s,\beta)$	-122,6	-37.0	1685~2273	4
$Si(s) + 2Cl_2(g) = SiCl_4(g)$	-660.2	-128.8	334(B)~1685(m)	4
$3Si(s) + 2N_2(g) = Si_3N_4(s,\alpha)$	-723.8	-315.1	298~1685(m)	4
$3Si(l) + 2N_2(g) = Si_3N_4(s,\alpha)$	-874.5	-405.0	1685(m)~1973	4
$Si(s) + \frac{1}{2}O_2(g) = SiO(g)$	-104.2	82.5	298~1685(m)	4
$Si(s) + O_2(g) = SiO_2\ (s,quartz)$	-907.1	-175.7	298~1685(m)	4
$Si(s) + O_2(g) = SiO_2(s,\beta\text{-}cristobalite)$	-904.8	-173.8	298~1685(m)	2
$Si(l) + O_2(g) = SiO_2(s,\beta\text{-}cristobalite)$	-946.3	-197.6	1685~1996(M)	2
$Si(l) + O_2(g) = SiO_2(l)$	-921.7	-185.9	1996~3514(b)	2
$Si(s) + \frac{1}{2}S_2(g) = SiS(g)$	51.8	81.6	973~1685(m)	2
$Si(s) + S_2(g) = SiS_2(s)$	-326.4	-139.0	298~1363(M)	4
$Sn(l) + Cl_2(g) = SnCl_2(l)$	-333.0	-118.4	520~925(B)	3
$Sn(l) + Cl_2(g) = SnCl_2(g)$	-225.9	-13.0	925(B)~1473	4
$Sn(l) + 2Cl_2(g) = SnCl_4(g)$	-512.5	-150.6	500~1200	3
$Sn(l) + \frac{1}{2}O_2(g) = SnO(g)$	6.3	50.9	505(m)~1973	
$Sn(l) + O_2(g) = SnO_2(s)$	-574.9	-198.4	505~1903(M)	
$Sn(l) + \frac{1}{2}S_2(g) = SnS_2(g)$	26.0	49.4	505~1973	8

Reaction	ΔH° (kJ/mol)	ΔS° (J/mol K)	Temperature Range (K)	Ref
$Ta(s) + \tfrac{1}{2}O_2(g) = TaO(g)$	188.0	87.0	298~2273	15
$Ta(s) + O_2(g) = TaO_2(g)$	-209.0	20.5	298~2273	15
$\tfrac{1}{2}Te_2(g) + \tfrac{1}{2}O_2(g) = TeO(l)$	-7.1	6.0	1282~1973	8
$Ti(s) + B(s) = TiB(s)$	-163.0	-5.9	298~1939	15
$Ti(s) + 2B(s) = TiB_2(s)$	-285.0	-20.5	298~1939	15
$Ti(s) + O_2(g) = TiO_2\ (s, rutile)$	-941.0	-177.6	298~1943(m)	4
$Ti(s) + 2Cl_2(g) = TiCl_4(g)$	-764.0	-121.5	298~1943(m)	4
$3Ti(s) + 2\tfrac{1}{2}O_2(g) = Ti_3O_5(s)$	-2,435.1	-420.5	298~1943(m)	4
$2Ti(s) + 1\tfrac{1}{2}O_2(g) = Ti_2O_3(s)$	-1,502.1	-258.1	298~1943(m)	4
$Ti(s) + \tfrac{1}{2}O_2(g) = TiO\ (s,\beta)$	-514.6	-74.1	298~1943(m)	4
$U(l) + C(s) = UC(s)$	-109.6	-1.8	1405(m)~1800	4,7
$U(s) + \tfrac{1}{2}N_2(s) = UN(s)$	-292.9	-80.8	298~1405(m)	4,7
$U(s) + O_2(g) = UO_2(s)$	-1,079.5	-167.4	298~1405(m)	3
$U(l) + O_2(g) = UO_2(s)$	-1,086.6	-172.3	1405~2273	4
$V(s) + B(s) = VB(s)$	-138.0	-5.9	298~2273	8
$V(s) + 0.23N_2(g) = VN_{0.23}(s)$	-130.0	-44.4	298~1973	15
$V(s) + \tfrac{1}{2}N_2(g) = VN(s)$	-214.6	-82.4	298~2619	4
$V(s) + O_2(g) = VO_2(s)$	-706.0	-155.0	298~1633(M)	15
$2V(s) + 1\tfrac{1}{2}O_2(g) = V_2O_3(s)$	-1,203.0	-238.0	298~2343	15
$V(s) + \tfrac{1}{2}O_2(g) = VO(s)$	-425.0	-80.0	298~2073	15
$W(s) + C(s) = WC(s)$	-42.3	-5.0	1173~1575	4
$2W(s) + C(s) = W_2C(s)$	-30.5	2.3	1575~1673	19
$Zn(l) + \tfrac{1}{2}S_2(g) = ZnS(s)$	-277.8	-107.9	693(m)~1180(b)	4,8
$Zn(g) + \tfrac{1}{2}S_2(g) = ZnS(s)$	-375.4	-191.6	1120~2000	3
$Zn(g) + \tfrac{1}{2}S_2(g) = ZnS(g)$	5.0	-30.5	1180~1973	4,8
$Zr(s) + 2B(s) = ZrB_2(s)$	-328.0	-23.4	298~2125	15
$Zr(s) + C(s) = ZrC(s)$	-196.6	-9.2	298~2125(m)	4,7
$Zr(s) + \tfrac{1}{2}N_2(g) = ZrN(s)$	-363.0	-92.0	298~2125(m)	4,7
$Zr(s) + \tfrac{1}{2}S_2(g) = ZrS(g)$	237.2	78.2	298~2125(m)	4,8

$Zr(s) + S_2(g) = ZrS_2(s)$	-698.7	-178.2	298~1823(M)	4,8

Note : (m) = melting point of metal, (M) = melting point of compound
(b) = boiling point of metal, (B) = boiling point of compound

References :

1. J.P. Coughlin, *Contributions to the Data on Theoretical Metallurgy. XII. Heats and Free Energies of Formation of Inorganic Oxides*, Bulletin 542, Bureau of Mines, U.S. Department of the Interior, Washington, D.C., 1954.
2. D.R. stull and H. prophet, *JANAF Thermochemical Tables*, 2nd ed., NSRDS-NBS 37, U.S. Department of Commerce, Washington, D.C., 1971.
3. O. Kubaschewski and C. B. Alcock, *Metallurgical Thermochemistry*, 5th ed., Pergamon Press, New York, 1979.
4. E. T. Turkdogan, *Physical Chemistry of High Temperature Technology*, Academic Press, New York, 1980.
5. H. H. Kellogg, *Trans. Met. Soc. AIME*, 236:602, 1966.
6. H. H. Kellogg, *J. Chem. Eng. Data*, 14:41, 1969.
7. I. Barin and O. Knacke, *Thermodynamical Properties of Inorganic Substances*, Springer-Verlag, New York, 1973.
8. I. Barin, O. Knacke, and O. Kubaschewski, *Thermodynamical Properties of Inorganic Substances*, Supplement, Springer-Verlag, New York, 1977.
9. M. Nagamori and T. R. Ingraham, *Metall. Trans.*, 1:1821, 1970.
10. H. H. Kellogg, *Metall. Trans.*, 2:2161, 1971.
11. G. J. Janz and G.M. Dijkhuis, in *Molten Salts*, vol.2, NBS, U.S. Department of Commerce, Washington, D.C., 1969.
12. H. Nagai, *J. Electrochem. Soc.*, 126:1400, 1979.
13. A. D. Mah and L.B. Pankratz, *Contributions to the Data on Theoretical Metallurgy, XVI. Thermodynamic properties of Nickel and Its Inorganic Compounds*, Bulletin 668, Bureau of Mines, U.S. Department of the Interior, Washington, D.C. 1976.
14. L. B. Pankratz, *Thermodynamic Properties of Halides*, Bulletin 674, Bureau of Mines, U.S. Department of the interior, Washington, D.C. 1984.
15. M.W. chase, *JANAF Thermochemical Tables*, 3rd ed.,1985, *American chem. Soc. And the American institute of physics for the national Bureau of standard. J. phys. Chem. Ref. Data Vol 14*, Supplement NO.1, Michigan 48674 USA, 1985.
16. Turkdogan, E. T. Rice, B. B., and Vinters, J. V. *Sulfide and Sulfate Solid Solubility in Lime, Metall. Trans.*, 5, 1527~35., 1974
17. Skeaff, J. M. and Espelund, A. W., *An E.M.F. Method for sulfate-oxide equilibria results for the Mg, Mn, Fe, Ni, Ca and Zn system.*, Can. Metall., Q., 12, 445~58., 1973
18. Kubaschewski, O., *The thermodynamic properties of double oxides High temperature – High pressure*, 4, 1~12., 1972.
19. Gupta, D. K. and seigle. L. L., *Free energies of formation of WC and W$_2$C, and the thermodynamic properties of carbon in solid tungsten*, Metall. Trans. A6, 1939~44,

1975.
20. Baker, F. B. and Holley, C. E., *Enthalpy of formation of cerium sesquioxide.* J. chem. Eng. Data, 13, 405~8, 1968.
21. Toker, N., ph. D., *Thesis.*, Pennsylvania state university, 1977.

APPENDIX III

Properties of Selected Elements

Atomic number	Element Symbol	Element Name	Atomic weight	Density* (kg m^{-3})	Melting point, K	Boiling point, K
1	H	Hydrogen	1.0079	(0.090)	14.025	20.268
2	He	Helium	4.00260	(0.179)	0.95	4.215
3	Li	Lithium	6.941	530	453.7	1615
4	Be	Beryllium	9.01218	1850	1560	2745
5	B	Boron	10.81	2340	2300	4275
6	C	Carbon	12.011	2620	4100	4470
7	N	Nitrogen	14.0067	(1.250)	63.14	77.35
8	O	Oxygen	15.9994	(1.429)	50.35	90.18
9	F	Fluorine	18.9984	(1.696)	53.48	84.95
10	Ne	Neon	20.179	(0.901)	24.553	27.096
11	Na	Sodium	22.9898	970	371.0	1156
12	Mg	Magnesium	24.305	1740	922	1363
13	Al	Aluminium	26.9815	2700	933.25	2793
14	Si	Silicon	28.0855	2330	1685	3540
15	P	Phosphorus	30.9738	1820	317.3	550
16	S	Sulphur	32.06	2070	388.36	717.75
17	Cl	Chlorine	35.453	(3.17)	172.16	239.1
18	Ar	Argon	39.948	(1.784)	83.81	87.30
19	K	Potassium	39.0983	860	336.35	1032
20	Ca	Calcium	40.08	1550	1112	1757
21	Sc	Scandium	44.9559	3000	1812	3104
22	Ti	Titanium	47.90	4500	1943	3562
23	V	Vanadium	50.9415	5800	2175	3682
24	Cr	Chromium	51.996	7190	2130	2945
25	Mn	Manganese	54.9380	7430	1517	2335
26	Fe	Iron	55.847	7860	1809	3135
27	Co	Cobalt	58.9332	8900	1768	3201
28	Ni	Nickel	58.70	8900	1726	3187

Atomic number	Element Symbol	Element Name	Atomic weight	Density* (kg m⁻³)	Melting point, K	Boiling point, K
29	Cu	Copper	63.546	8960	1357.6	2836
30	Zn	Zinc	65.38	7140	692.73	1180
31	Ga	Gallium	69.72	5910	302.90	2478
32	Ge	Germanium	72.59	5320	1210.4	3107
33	As	Arsenic	74.9216	5720		876
34	Se	Selenium	78.96	4800	494	958
35	Br	Bromine	79.904	3120	265.9	332.25
36	Kr	Krypton	83.80	(3.74)	115.78	119.80
37	Rb	Rubidium	85.4678	1530	312.64	961
38	Sr	Strontium	87.62	2600	1041	1650
39	Y	Yttrium	88.9059	4500	1799	3611
40	Zr	Zirconium	91.22	6490	2125	4682
41	Nb	Niobium	92.9064	8550	2740	5017
42	Mo	Molybdenum	95.94	10200	2890	4912
43	Tc	Technetium	98	11500	2473	4538
44	Ru	Ruthenium	101.07	12200	2523	4423
45	Rh	Rhodium	102.9055	12400	2236	3970
46	Pd	Palladium	106.4	12000	1825	3237
47	Ag	Silver	107.868	10500	1234	2436
48	Cd	Cadmium	112.41	8650	594.18	1040
49	In	Indium	114.82	7310	429.76	2346
50	Sn	Tin	118.69	7300	505.06	2876
51	Sb	Antimony	121.75	6680	904	1860
52	Te	Tellurium	127.60	6240	722.65	1261
53	I	Iodine	126.9045	4920	386.7	458.4
54	Xe	Xenon	131.30	(5.89)	161.36	165.03
55	Cs	Cesium	132.9054	1870	301.55	944
56	Ba	Barium	137.33	3500	1002	2171
57	La	Lanthanum	138.9055	6700	1193	3730
58	Ce	Cerium	140.12	6780	1071	3699
60	Nd	Neodymium	144.24	7000	1289	3341

Atomic number	Element Symbol	Element Name	Atomic weight	Density* (kg m⁻³)	Melting point, K	Boiling point, K
62	Sm	Samarium	150.4	7540	1345	2064
72	Hf	Hafnium	178.49	13100	2500	4876
73	Ta	Tantalum	180.9479	16600	3287	5731
74	W	Tungsten	183.85	19300	3680	5828
75	Re	Rhenium	186.207	21000	3453	5869
76	Os	Osmium	190.2	22400	3300	5285
77	Ir	Iridium	192.22	22500	2716	4701
78	Pt	Platinum	195.09	21400	2045	4100
79	Au	Gold	196.9665	19300	1337.58	3130
80	Hg	Mercury	200.59	13530	234.28	630
81	Tl	Thallium	204.37	11850	577	1746
82	Pb	Lead	207.2	11400	600.6	2023
83	Bi	Bismuth	208.9804	9800	544.52	1837
84	Po	Polonium	209	9400	527	1235
86	Rn	Radon	222	(9.91)	202	610
88	Ra	Radium	226.0254	5000	973	1809
89	Ac	Actium	227.0278	10070	1323	3473
90	Th	Thorium	232.0381	11700	2028	5061
92	U	Uranium	238.029	18900	1405	4407
94	Pu	Plutonium	244	19800	913	3503

* Density at 300K for solids and liquids, and at 273K for gases indicated by ( ).

APPENDIX IV

Standard half-cell potentials in aqueous solutions

(T = 298 K, Standard state = 1 molal)

Electrode reaction	Potentials (V)	
	Oxidation	Reduction
Acid solutions		
$Li = Li^+ + e$	3.045	-3.045
$K = K^+ + e$	2.925	-2.925
$Cs = Cs^+ + e$	2.923	-2.923
$Ba = Ba^{2+} + 2e$	2.90	-2.90
$Ca = Ca^{2+} + 2e$	2.87	-2.87
$Na = Na^+ + e$	2.714	-2.714
$La = La^{3+} + 3e$	2.52	-2.52
$Mg = Mg^{2+} + 2e$	2.37	-2.37
$2H^- = H_2 + 2e$	2.25	-2.25
$Th = Th^{4+} + 4e$	1.90	-1.90
$U = U^{3+} + 3e$	1.80	-1.80
$Al = Al^{3+} + 3e$	1.66	-1.66
$Mn = Mn^{2+} + 2e$	1.18	-1.18
$Zn = Zn^{2+} + 2e$	0.763	-0.763
$Cr = Cr^{3+} + 3e$	0.74	-0.74
$U^{3+} = U^{4+} + e$	0.61	-0.61
$O_2^- = O_2 + e$	0.56	-0.56
$S^{2-} = S + 2e$	0.48	-0.48
$Ni + 6NH_3(aq) = Ni(NH_3)_6^{2+} + 2e$	0.47	-0.47
$Fe = Fe^{2+} + 2e$	0.44	-0.44
$Cu + CN^- = CuCN_2^- + e$	0.43	-0.43
$Cr^{2+} = Cr^{3+} + e$	0.41	-0.41
$Cd = Cd^{2+} + 2e$	0.403	-0.403
$Pb + SO_4^{2-} = PbSO_4 + 2e$	0.356	-0.356

Electrode reaction	Potentials (V)	
	Oxidation	Reduction
$Tl = Tl^+ + e$	0.336	-0.336
$Co = Co^{2+} + 2e$	0.277	-0.277
$Pb + 2Cl^- = PbCl_2 + 2e$	0.268	-0.268
$Ni = Ni^{2+} + 2e$	0.250	-0.250
$Ag + I^- = AgI + e$	0.151	-0.151
$Sn = Sn^{2+} + 2e$	0.136	-0.136
$Pb = Pb^{2+} + 2e$	0.126	-0.126
$Cu + 2NH_3(aq) = Cu(NH_3)_2^+ + e$	0.12	-0.12
$H_2 = 2H^+ + 2e$	0.000	0.000
$2S_2O_3^{2-} = S_4O_6^{2-} + 2e$	-0.08	0.08
$Ag + Br^- = AgBr + e$	-0.095	0.095
$H_2S = S + 2H^+ + e$	-0.141	0.141
$Sn = Sn^{2+} + 2e$	-0.15	0.15
$Ag + Cl^- = AgCl + e$	-0.222	0.222
$2Hg + 2Cl^- = Hg_2Cl_2 + 2e$	-0.2677	0.2677
$Cu = Cu^{2+} + 2e$	-0.337	0.337
$Cu = Cu^+ + e$	-0.521	0.521
$2I^- = I_2 + 2e$	-0.5355	0.5355
$Fe^{2+} = Fe^{3+} + e$	-0.771	0.771
$2Hg = Hg_2^{2+} + 2e$	-0.789	0.789
$Ag = Ag^+ + e$	-0.7991	0.7991
$Hg_2^{2+} = 2Hg^{2+} + 2e$	-0.920	0.920
$Au + 4Cl^- = AuCl_4^- + 3e$	-1.0	1.0
$2Br^- = Br_2(l) + 2e$	-1.065	1.065
$H_2O = \frac{1}{2}O_2 + 2H^+ + 2e$	-1.229	1.229
$Mn^{2+} + 2H_2O = MnO_2 + 4H^+ + 2e$	-1.23	1.23
$2Cr^{3+} + 7H_2O = Cr_2O_7^{2-} + 14H^+ + 6e$	-1.33	1.33
$2Cl^- = Cl_2 + 2e$	-1.3595	1.3595
$Pb^{2+} + 2H_2O = PbO_2 + +4H^+ + 2e$	-1.455	1.455

Electrode reaction	Potentials (V)	
	Oxidation	Reduction
$Ce^{3+} = Ce^{4+} + e$	-1.61	1.61
$PbSO_4 + 2H_2O = PbO_2 + SO_4^{2-} + 4H^+ + 2e$	-1.685	1.685
$Co^{2+} = Co^{3+} + e$	-1.82	1.82
$Ag^+ = Ag^{2+} + e$	-1.98	1.98
$2SO_4^{2-} = S_2O_8^{2-} + 2e$	-2.01	2.01
$O_2 + H_2O = O_3 + 2H^+ + 2e$	-2.07	2.07
$2F^- = F_2 + 2e$	-2.89	2.89
$SO_4^{2-} = S(s) + 2O_2 + 2e$	-3.8587	3.8587
Basic solutions		
$Ca + 2OH^- = Ca(OH)_2 + 2e$	3.03	-3.03
$Cr + 3OH^- = Cr(OH)_3 + 3e$	1.3	-1.3
$Zn + 4OH^- = ZnO_2^{2-} + 2H_2O + 2e$	1.126	-1.126
$CN^- + 2OH^- = CNO^- + H_2O + 2e$	0.97	-0.97
$SO_3^{2-} + 2OH^- = SO_4^{2-} + H_2O + 2e$	0.93	-0.93
$Fe + 2OH^- = Fe(OH)_2 + 2e$	0.877	-0.877
$H_2 + 2OH^- = 2H_2O + 2e$	0.828	-0.828
$Ni + 2OH^- = Ni(OH)_2 + 2e$	0.72	-0.72
$Fe(OH)_2 + OH^- = Fe(OH)_3 + e$	0.56	-.56
$2Cu + 2OH^- = Cu_2O + H_2O + 2e$	0.358	-0.358
$2Ag + 2OH_- = Ag_2O + H_2O + 2e$	-0.344	0.344
$4OH^- = O_2 + 2H_2O + 4e$	-0.401	0.401
$Hg_2^{2+} = 2Hg^{2+} + 2e$	-0.920	0.920
$O_2 + 2OH^- = O_3 + H_2O + 2e$	-1.24	1.24

INDEX

Activity
    activity 67
    aqueous solution 260
    coefficient 71, 83
    coefficient, mean ionic 262
    mean, ionic 261
    quotient 97
Adiabatic expansion 29, 32
Adiabatic compression 29, 32
Adiabatic flame temperature 55
Alkemade line 213
Alkemade theorem 214
Alkemade triangle 214
Allotrope 163
Allotropy 163
Alpha function 91
Anode 233
Aqueous solution 260
Avogadro's number 233
Azeotrope 172

Binary
    eutectic 178
    solution 167
    system 166, 186
    system, with solid solution 186
    system, without solid solution 177
Bivariant 210
Boiling
    boiling 161
    temperature 137
Boltzmann's constant 46
Boundary curve 212

Carnot cycle 28
Cathode 233
Cell
    chemical 251
    concentration 251
    Daniell 244
    electrochemical 243
    electrode concentration 251
    electrolyte concentration 251, 255
        with transference 251, 255
        without transference 251
    galvanic 233, 243
Chemical cell 251
Chemical potential 63, 67, 69
Clapeyron equation 139
Clausius- Clapeyron equation 140
Clustering 86
Performance coefficient of heat pump 31
Common tangent 147
Component 132
Composition triangle 204
Compound formation 86
Compressibility
    compressibility 42
    factor 65
Compression
    adiabatic 29,32
    isothermal 29, 32
Concentration cell 251
Congruent melting 180, 191, 212
Conjugate phases 228
Conjugation line 214
Cooling curve 161
Coring 193
Criterion of equilibrium 101, 271
Critical pressure 159, 162
Critical temperature
    critical temperature 153, 159, 162
    lower 173
    upper 173
Crystallisation path 214-222

Daniell cell 244
Debye-Huckel limiting law 264
Degradation 17
Degree of freedom 131, 134
Degree of irreversibility 20
Dilute solution 87
Distillation 175

Effective concentration 71
Efficiency of engine 29
Electrochemical cell 243
Electrochemical equilibrium 235
Electrochemical reaction 234
Electrochemistry 233
Electrode
    electrode 243
    hydrogen 243
    reference 246
    standard hydrogen 246
Electrolysis 241
Electrolyte
    electrolyte 245
    solid 253
Electromotive force 238
Ellingham diagram 127
Emf 238
Endothermic 17, 52
Energy
    energy 1
    chemical 1
    electrical 1
    mechanical 1
Enthalpy
    enthalpy 6
    change 13
    of formation 49
    of reaction 51
Entropy
    entropy 17, 21
    change 21
    absolute 45
Equilibrium
    equilibrium 16, 27
    constant 97, 98
    criteria 27, 101
    effect of pressure 114
    effect of temperature 111
    electrochemical 235
    phase 131
    reaction 97
Eutectic
    binary 177
    point 178
    reaction 179
    structure 188
    system 177
    ternary 214
Eutectoid
    point 188
    reaction 188
Excess integral molar free energy 85
Excess molar free energy of mixing 85
Excess partial molar enthalpy 85
Excess partial molar entropy 85
Excess partial molar free energy 85
Excess property 85
Exothermic 17, 52
Expansion
    adiabatic 29, 32
    coefficient 42
    isothermal 29, 32
Extensive property 4
Extremum principle 101
Faraday's law 233

First law of thermodynamics 1
Fractional distillation 175
Free energy
    free energy 32, 33
    change 33, 34
    minimisation method 104, 108
    Gibbs 34, 57
    Helmholtz 33
    effect of pressure 37
    effect of temperature 38
    ion 260
Fugacity 64, 69
Fusion curve 158

Galvanic cell 233, 243
Gibbs free energy
    Gibbs free energy 34, 57
    change 34
    of mixing 62
Gibbs phase rule 134
Gibbs-Duhem equation 76, 89
Gibbs-Helmholtz equations 39

Half-cell 233
Heat 1
Heat capacity
    heat capacity 7
    constant pressure 7
    constant volume 7
Heat engine 28
Heat of formation 49
Heat of fusion 14
Heat of reaction 51
Heat pump 29
Heat reservoir 28, 30
Heating path 219
Henry's law 88
Hess's law 15
Hydrogen electrode 243
Hypereutectic 188
Hypoeutectic 188

Ideal gas 61
Ideal gas equation 61
Ideal solution 80, 81, 83
Incongruent melting 181, 191, 213
Integral molar volume 76
Integral property 76
Intensive property 4
Interaction coefficient 125
Interaction parameter 93
Intermediate phase 154
Internal energy 1
Invariant 184, 210
Irreversibility 20
Irreversible process 16
Isobar 157

Isotherm
    isotherm 157
    liquidus 212
Isothermal plane 208
Isothermal compression 29, 32
Isothermal expension 29, 32
Isothermal section 223

Join 214
Junction potential 255

Kirchhoff's law 53

Le Chatelier's principle 116
Lever rule 148
Liquid solution 144
Liquidus 171, 178
Liquidus isotherm 212

Margules equation
    Margules equation 94
    three suffix 95
Maxwell's equations 41
Mean activity coefficient 262
Mean activity of ion 262
Mean ionic molality 262
Mean molar free energy of ion 261
Mean thermodynamic property 261
Melting temperature 137
Melting
    congruent 180, 191, 212
    incongruent 181, 191, 213
Meta-stable phase 155
Miscibility gap 148
Miscibility
    miscibility 166, 172
    partial 172
    total 166
Molality 261
Molar free energy of mixing 79
Molarity 269
Monotectic
    point 183
    reaction 183

system 183

Natural process 16
Negative deviation 83
Nernst equation 239
Non-equilibrium state 16, 27
Non-ideal solution 83
Non-state function 4

Order-disorder transition 193
Ordered structure 193
Oxidation 233
Oxygen sensor 253

Partial molar energy 74
Partial molar enthalpy 74
Partial molar entropy 74
Partial molar free energy
    partial molar free energy 63, 74
    of mixing 78
    mean 261
Partial molar volume 74, 75
Partial property 74
Perfect gas 61
Perfect gas equation 61
Peritectic
    point 181
    reaction 182
    system 181
    temperature 181
    ternary 214
pH 269
Phase
    phase 132
    boundary 139, 157
    diagram 150, 157, 178
    equilibrium 131
    intermediate 154
    meta-stable 155
    primary 212
    reaction 179
    separation 86, 147
    terminal 151, 154
    transformation 136
    transition 138
Phase rule 131, 134
Polymorphic transformation 163
Polymorph 163
Polythermal projection 211
Positive deviation 83
Pressure-Temperature diagram 157
Primary field 178, 212
Primary phase 212

Raoult's law 81
Real gas 64
Reduction 233
Reference electrode 246
Reference state 47
Regular solution 93
Relative integral molar free energy 79
Relative molar free energy 61
Relative partial molar free energy 78
Reversible process 16, 17
Reversible work 3

Second law of thermodynamics 16
Solid electrolyte 253
Solid solubility
    solid solubility 186
    partial 187
    total 186
Solid solution
    intermediate 191
    interstitial 186
    substitutional 186
    terminal 191, 226
Solidus 178
Solubility constant 268
Solubility product 268
Solution model 92
Solution
    solution 61
    aqueous 260
    dilute 87
    ideal 80, 81, 83
    liquid 144
    model 92

non-ideal 83
regular 93
solid 144
Spontaneous process 16
Standard electrode potential 247
Standard free energy of formation 58
Standard half-cell potential 247
Standard hydrogen electrode 246
Standard state
   standard state 47
   alternative 117
   1 molality 262
   1wt% 120
   Henrian 118
   infinitely dilute solution 120
   pure state 118
     Raoultian 118
State function 4
Sublimation curve 158
Surface tension 1
Surroundings 1, 2
System
   system 1, 2
   binary 166, 186
   closed 75
   eutectic 177
   monotectic 183
   multicomponent 77
   open 75
   peritectic 181
   ternary 204
   unary 157

Temperature coefficient 239
Temperature-composition diagram 171
Terminal phase 154

Terminal solid solution 191
Ternary eutectic 214
Ternary peritectic 214
Ternary system 204
Theoretical plate 175
Thermodynamic model
   ideal solution 195
   non-ideal solution 198
   regular solution 200
Third law of thermodynamics 43
Tie line 178, 228
Tie triangle 229
Transition temperature 138
Triple point 135, 139
Trivariant 210

Unary system 157
Univariant 185, 210

van der Waals equation 67
van't Hoff equation 119
Vaporisation curve 158
Vaporus 171
Virial coefficient 65

Work
   work 1
   additional 34, 238
   electrical 1
   irreversible 19
   magnetic 1
   maximum 19
   mechanical 1
   PV 33
   reversible 3, 19